A Place in the Sun

A Place in the Sun

Shetland and Oil—Myths and Realities

Jonathan Wills

Social and Economic Studies No. 41
Institute of Social and Economic Research
Memorial University of Newfoundland

Published by
The Institute of Social and Economic Research
Memorial University of Newfoundland
St. John's, Newfoundland
Canada
ISBN 0-919666-69-8

Printed on paper
containing over 50%
recycled paper including
5% post-consumer fibre.

Canadian Cataloguing in Publication Data

Wills, Jonathan

A place in the sun

(Social and economic studies, ISSN 0847-0898 ;
no. 41)

ISBN 0-919666-69-8

1. Offshore oil industry -- Scotland -- Shetland --
Economic aspects. 2. Offshore oil industry --
Scotland -- Shetland -- Social aspects.
3. Shetland (Scotland) -- Economic conditions.
4. Shetland (Scotland) -- Social conditions.
I. Memorial University of Newfoundland. Institute
of Social and Economic Research. II. Title.
III. Series: Social and economic studies
(St. John's, Nfld.) ; no. 41.

This book is dedicated to my friends in Cordova District Fishermen United, Prince William Sound, Alaska, whose spirit in adversity has been an inspiration.

This is what £1.3 billion looks like: Sullom Voe terminal in its completed state, with all four jetties in use.
Photo: BP

Contents

List of Maps

List of Diagrams

Acknowledgements

My publishers have been very patient. This book was supposed to be ready years ago but always there was some new excuse: the plot was still thickening, with no dénouement in sight; I was very busy at work; or another very welcome small person had come to live in our household, disrupting my sleep with bawling and nappies and making serious work on this book impossible for many months. If ISER Books had been based in Scotland they would have sent round the heavy mob long ago to hurry me up. The Atlantic Ocean protected me. But here we are at last and I must record a debt of gratitude to my editors, Robert Paine and Susan G. Nichol (and to her equally charming predecessors who never gave up hope that I would produce a manuscript in the end) and to Jeanette Gleeson of ISER Books and Memorial University of Newfoundland.

The Shetland Arts Trust, who have sponsored many a worthier cause, were kind enough to make me a six-month bridging loan to purchase the electronic ironmongery upon which this book has been written and without which it would have been even worse written.

I am grateful also to the many professional colleagues in Scottish journalism who have discussed the themes of this book with me at various times over the past 17 years, not least to my long-suffering fellow workers at *The Shetland Times*; to Basil Wishart, the former editor of that newspaper who, back in 1972, had the foresight to open a cuttings file on the oil industry in Shetland—an invaluable work of reference of which I have no doubt made inadequate use; to Bertie and Ena Mathieson of Noosthamar, Buddabreck, Unst, whose generous hospitality enabled me to make a start on the first draft; to Chris Baur, who researched the Nordport land speculation in Shetland and encouraged me in this work; to my early collaborator, Ron MacKay, who urged me, with little success, to adopt a more methodical and scientific approach to my subject; to various members of the Labour Party, the Scottish National Party and the Shetland Movement who discussed oil and Shetland with me; to those much-maligned public relations men from BP—Mike Whittall, Ken Welsh, Philip Wyper, Peter Wyllie, Peter Johnson, Peter Guy and, above all, Erik Arthur, who put their company's case with skill and energy, told me as much as they were able, arranged interviews with members of BP management, patiently explained all sorts of technical matters and never took it personally when I bit the hand that fed me; to Captains George Biro, Bert Flett and George Sutherland (not forgetting Jim Dickson, the computer king at SIC pollution control) who made sure that I usually got an answer to my exasperating questions at Shetland Islands Council's port control office; to many other council officials for being patient, courteous and helpful with my inquiries; and in particular to Ian R. Clark, the former SIC chief executive who gave me straight answers to straight questions—Shetland owes him a great deal and I hope that this book will help more people to understand why.

Many elected members of Shetland Islands Council have been helpful to me over the years. Some of them have been extremely hospitable, willing to debate oily local politics well into the small hours over their own whisky. Those were the days.

I wish I could move a vote of thanks to the council members, collectively, for unstinting assistance in providing me with the full facts about the council's relationship with the oil industry. Alas, I cannot. The door of Lerwick Town Hall has often been closed in recent years to reporters asking questions which, on the other side of the Atlantic, would be answered more frankly and with less hesitation. I regret that sometimes friendship has decayed into wary acquaintance because of the differences of opinion between us but I hope that this book will help them to understand my annoying point of view a little better.

My numerous "moles" and "whistleblowers" in the Sullom Voe terminal and Shetland Islands Council would not wish to be acknowledged by name but I must record my appreciation of the dangerous work they do for no reward—and for the plain brown envelopes with interesting enclosures which continue to arrive on my desk. You're doing a grand job, lads. Keep it up!

For help with the chapter on Alaskan parallels I am indebted to Harry Evans, formerly my editor on the *London Times* and now editor-in-chief of *Condé Nast Traveler* in New York, who sent me to Alaska to cover the *Exxon Valdez* story; to Rick Steiner, David Grimes, Riki Ott, Marilyn Leland, Cliff Ward, Jerry McCune, Paula

Lamb, Janine Buller and others too numerous to mention in Cordova District Fishermen United; to Charles J. Hamel in Alexandria, Virginia; to Tom Brennan of Alyeska Pipeline Service Company, Jack Hallerin of Exxon, Commander Steve McCall of the US Coast Guard and Commander Jim Woodle, US Coast Guard (rtd), all of whom answered my awkward questions courteously; to Jon Rush of Windsinger Haven, Ellamar, Alaska; and to that admirable and wise fellow, Dan Lawn of the Alaska State Department of Environmental Conservation in Valdez, who was kind enough to check my manuscript during his busy visit to Shetland in September 1989 and spared me the embarrassment of printing even more careless mistakes about his adopted home.

The endless re-draftings were done to the endlessly-repeated strains of Bob Dylan's only flawless album, *The Freewheelin' Bob Dylan*. If that is a useful hint to fellow labourers in this thin vinyard, so be it.

Lastly, and most importantly, my wife, Lesley Roberts, has been an honest critic of my drafts (if a bit outspoken at times!)—with a greater regard for the truth than for my self-esteem. She has tolerated many sacrifices in our domestic life for the sake of this book and I may not have made clear to her how much I appreciated that. For her forbearance and loyalty, and much else, I thank her. It will be nice to spend a few quiet winter evenings together again, while the word processor gathers some well-earned dust.

Jonathan Wills
Bressay, Shetland

September 1990

Uyeasound, Unst: where the advance guard of the “hippy commune” landed in May 1973.
Photo: Jonathan Wills

Introduction
"A Hippy on Social Security"

It was 1972. Everyone who could handle a typewriter was rushing out an instant book on North Sea oil. Me too. When the London publisher Tom Stacey asked me to write "Shetland's Oil", with my friend and university colleague, Ron MacKay, I was flattered, enthusiastic and, it must be admitted, eager to get my hands on some of Mr Stacey's cash.

There was much need of the money; with five fellow graduates from Edinburgh University, I was about to start a mad project in the Shetland Islands, bringing 27 acres of neglected crofting land and two derelict houses back into productive use—in Haroldswick, at the north end of Unst, Britain's most northerly island. Haroldswick is a few miles north and 2,000 miles east of Cape Farewell, the southern tip of Greenland, with a wild maritime climate which rivals the bleaker bits of Newfoundland for gales, rain and occasional "days atween wadders"—that is, days between weathers which are the only reason anyone stays there.

Within four years the project was accomplished, after much labour and considerable domestic upheavals which need not detain us here but may one day make a posthumous tragi-comic novel. The mad bit was that we had decided to do this on the eve of the Shetland oil age—which was to change, for a while, everyone's assumptions about the traditional Shetland "way of life". And we planned to support ourselves entirely from our own efforts without claiming welfare payments from the state. We reasoned that we had no moral right to claim social security, although our financial circumstances would certainly have entitled us to considerable sums, because we could very easily support ourselves from trades other than subsistence farming—crofting, as they call it in the Highlands and Islands of Scotland. Indeed, we were all, in our different ways, to do just that when we eventually left Unst.

The crew of the Muckle Flugga attending boat *Grace Darling*: (from left to right) Willie Mathewson; Tom Bruce; Peter Sinclair; Bertie Mathieson; Johnnie Charlie Mathewson.
Photo: Jonathan Wills

The bitter end of Britain: the Muckle Flugga lighthouse.
Photo: Dennis Coutts

So we did not claim a cent during 1973 to 1976, nor ever afterwards, I am glad to say. We need not have been so high-minded. Many others in North Unst were receiving welfare payments, quite legally, and they would not or could not believe that we were not also on welfare. And so it was that, when I twice stood for Parliament as the Labour Party candidate for Orkney and Shetland, in February and October 1974, it was put about by the Liberal candidate's proxies that I was "a hippy living on social security". This hurt, particularly as I was a former Young Liberal and did not realise how dirty these people could fight. It also lost me my deposit in the October election—my only claim to political notoriety. The indignant efforts of the Unst representative of the Department of Health and Social Security, a kindly man who wished to set the record straight, were disregarded.

As I said, the money mattered. Along with my friends in the co-operative at Haroldswick, I earned it from odd jobs—some very odd indeed. There was crewing in the supply boat which took men and materials, including fresh water, out to the Muckle Flugga lighthouse—the last light at the bitter end of Britain. There was also selling some croft produce; fixing roofs and labouring work for neighbouring crofters, mostly for payment in kind; digging ditches for the telephone company; bits and pieces of journalism; writing books for children; selling a few paintings; and, above all, I put great faith in Mr Stacey's written promise of an advance on royalties, to be paid on delivery of my manuscript.

Writing the book had to be done in the evenings. In the winter, north of 60 degrees of latitude, the evenings are long, cold and dark. We had no electricity, only a Tilley paraffin pressure lamp, candles and peat fires that were never warm enough. It was difficult to write in the evenings, because our houses were small and shared. The noise of the typewriter woke up the babies who had joined us by this time. So Bertie Mathieson and his wife Ena took pity on me and offered the use of their spare room. Each evening, when I arrived on my pushbike with my papers in a backpack, they plied me with huge meals at their croft house of Noosthamar, overlooking the spectacular fjord of Burrafirth; they let me use all the electric light I wanted and were generally the most friendly and generous patrons a struggling author ever had.

Their efforts, alas, were to bear no fruit for 17 years. The bulk of the manuscript was indeed delivered to Mr Stacey but, in the very week when he was due to pay me my small advance, he went out of business. I never saw a penny of it and was pained, some years later, to hear that the gentleman's fortunes had revived and he was earning a handsome crust in some new commercial incarnation.

In disgust, I filed the book in an old tea chest and, by 1976, had taken up full-time journalism, a trade which has occupied me ever since. But the nights at Noosthamar, Old Norse for "the haven under the rocky slope", were not wasted, despite my first unfortunate experience of publishers. When, more than 10 years later, Robert Paine of ISER Books asked me to write this volume, I retrieved the mouldy manuscript from the tea chest and discovered that Mr Stacey's temporary financial embarrassment had spared me the embarrassment of publishing the most half-cocked, naive and foolish book ever written on North Sea oil, on Shetland or on any other topic. Only my colleague Ron MacKay's sections, on the technology and ecology of North Sea oil, had stood the test of time, although Ron had long since decamped to the USA, to follow a career as a research scientist.

Besides, even if I could have written it, 1972 was far too early to come out with a book about Shetland and oil. We had only the barest outline of what would become a very complicated plot. All we had to go on were consultants' reports, planners' proposals, environmentalists' scare stories and the beginnings of an attempt by local councillors in Shetland to bring the international oil industry under control in a very small place. Even now, in late 1990, we are only nearing the end of the second decade of a North Sea industry which seems certain to last for another three or four decades.

On 10th May 1973, I and the other member of the advance guard of the Haroldswick "hippy commune", as the locals insisted on calling it, landed in Unst from the Shetland inter-island steamer *Earl of Zetland*. We intended to prove that the "Shetland way of life" which we so admired was a sensible way to make a living in the last quarter of the 20th century. Already, most of Shetland had decided that that way of life was over; the future lay in making what they could out of "oil". They did so with energy, deserting their miserly and impecunious traditional local employers like lemmings, in the greatest economic emancipation of the Shetlanders since they were freed from debt bondage to their landlords by the Crofters Act of the 1880s.

No wonder they thought us mad young idealists, wasting our time (although few were rude enough to say it to our faces, at least when sober). But now, as we contemplate the slow decline of oil production over the coming decades, and the construction boom at Sullom Voe is just a memory, it is comforting for a middle-aged hippy to find that the values and lifestyle of pre-oil Shetland are once more being admired by the people who rejected them during the truly mad years of Shetland's oil boom. Many of us suspect that we will all, one day soon, have to return to the ecological principles which the Haroldswick "hippy commune" held so dear, back in the days when the Green Revolution was thought to be something that only concerned the more prosperous rice farmers of South East Asia, and the big companies, including some oil companies we were to get to know quite well, who sold them all that artificial fertiliser.

The way of life in Shetland after the oil boom will be very different from our little struggles to pay our way in Haroldswick and Burrafirth in 1973–76; there will be electricity, of course; and good roads, linked by the network of roll-on/roll-off car ferries, completed about the time we left Unst; and airplanes and helicopters for mundane transport and midnight medical emergencies, an important consideration if we are to retain population in remote islands many miles from a hospital; and fine new schools; and a system of social welfare which is already the envy of Britain. But this way of life will have to be paid for out of the renewable resources of the islands—peat for fuel; lamb, mutton, beef, poultry and fish for food; local stone for building. In the year 2025, the oil industry will still be in Shetland but it will be a dwindling trade—unless it turns out that the oilmen have told us even less than we suspect about what really lies under the seabed off this peculiar archipelago.

If there is a second oil bonanza it will only postpone the day when Mother Nature's chickens come home to roost. That will be a day of reckoning and no mistake. I hope that a few octogenarian hippies will still be around to see it and to cackle into their beards, whether or not they have compromised, in their dotage, and finally accepted the social security payments pressed upon them by a grateful local government.

That crofting co-operative, by the way, did not make any money. But it didn't lose any, either. The books balanced and the land and the buildings are still being put to better use than when I first saw them in July 1972. So there!

Since 1972 the simple story of Shetland and oil, of gallant little David and big bad Goliath, has developed a bewildering variety of bizarre sub-plots, with wildly improbable new characters, crazy scene shifts, unpredicted dénouements and several changes in the main script-writing team. It would all make a fine soap opera, or perhaps a real opera, but all I can offer is my own imperfect and thoroughly partial view of events. It is a small thing, but mine own.

Map of Shetland Islands

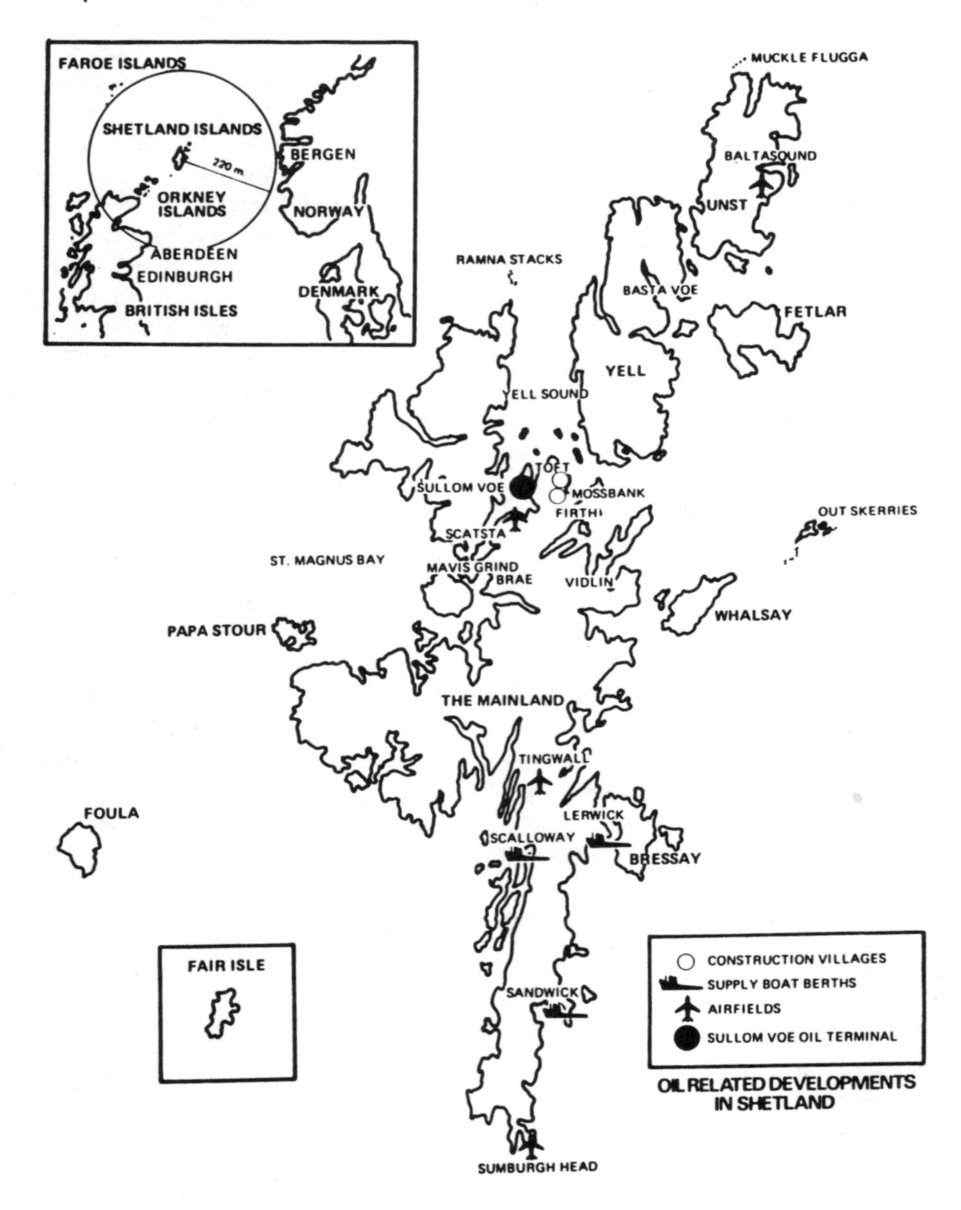

Credit: Reproduced from *Shetland's Oil Era* by Elizabeth Marshall with the permission of the publishers, The Research and Development Department of Shetland Islands Council.

Hornswoggle County 1

> The irony of the North Sea allegedly was that the tyrant which bestrode it was the Shetland County Council [sic], a tiny group of home-spun farmers led by 'Fuehrer' Ian Clark (Scot) who had reportedly hornswoggled some of the biggest multinationals and most sophisticated leaders in Britain out of terms which would make Scottish Nationalists pale with envy.
>
> Not only had these 'amateurs' pushed a private bill through Parliament, giving them reportedly unprecedented autonomy to make their own critical industrial decisions, but had forced the majors into giving them a highly profitable partnership for which they had put up no funds and over which they won significant control....
>
> Richard Funkhauser,
> United States Consul,
> Edinburgh, 1975.

The verb 'to hornswoggle' is thought to be of American origin. It means "to bamboozle" or "to swindle". This book is a belated attempt to find out just who was hornswoggled in Shetland between 1972 and 28th November 1988, when the 30 or so oil company partners in the giant oil terminal at Sullom Voe finally buried the hatchet with Shetland Islands Council after 16 years of increasingly bitter argument about money.

It is an interesting question, very interesting to other communities who have taken in, and taken on, the oil majors. To answer it, we must first examine the strange history of the decidedly odd community of 17,500 souls who, by 1972, lived in this 550 square mile archipelago, on the same latitude as Cape Farewell on the southern tip of Greenland. Many more of these bleak and sporadically beautiful islands were inhabited in the mid 19th century when they supported, after a fashion, over 30,000 people. Only 16 of the hundred islands of Shetland are inhabited today. There are plenty of good history books to tell you the details but the basic problem was that no-one except the occasional crazed hippy wanted to live without modern conveniences, above all electricity. And electricity in Shetland comes from diesel-powered generators, as it has always done since the island's only town, Lerwick, first switched on in 1936. The diesel fuel arrives in small coastal tankers from the BP refinery at Grangemouth on the River Forth, 400 miles to the south. For although the oil companies have built Europe's biggest oil terminal at Sullom Voe in Shetland, the oil still has to go south to be turned into useful things like diesel, heavy fuel oil, lubricants and plastics.

The little tankers dock most weeks at the storage tanks in Lerwick. Lerwick, now boasting about 8,000 inhabitants, styles itself the capital, although in truth it is just a county town, despite the greatly increased powers now delegated by the Westminster Parliament to Shetland Islands Council. The council meets in Lerwick Town Hall, a splendid Victorian Gothic building looking across to the island of Bressay which was the administrative centre of Shetland even before Scalloway, in the age of the Viking Earls, and which conveniently provides shelter for the commodious harbour.

The Town Hall is not really the centre of Lerwick. The centre is the Pierhead, and it always has been, ever since the enterprising Victorian businessmen of the town got together and set up a harbour trust to construct the first proper docks in Shetland. The life of Lerwick, and of much of Shetland, still centres on the bustling port with its three miles of quays, jetties, piers and oilfield supply bases. It is more of a North Atlantic crossroads than a remote British backwater. English is just one of the languages you may hear any morning along the Lerwick Pierhead. The Pierhead misses very little. The harbour is thronged each week with dozens of oil-related ships, in addition to passenger cargo ferries, naval ships, freighters and fishing vessels from Bulgaria to Greenland. Most arrivals are given the once-over. Numbers of elderly men and unoccupied youths (it is mostly a male pastime), patrol the waterfront sizing up the fleet and making loud and pointed remarks about ship design, the standard of housekeeping and maintenance on board, and the likely purpose of any 'stranger' vessel's visit to this marshalling yard of the north.

This has been going on at the Lerwick side of the water for 400 years and on the Bressay side for even longer. Seventeenth century Dutch herring boats were as closely examined by the Lerwick foreshore pundits as the first Viking longships had been by the Bressay Picts nine centuries before. Later, the vaunted nautical expertise of the "Pierhead Skippers" was employed to inspect Dutch East Indiamen, English fishing wherries, Dundee whalers, Royal

Navy men o' war, Press Gang tenders, sailing packets, Faroe cod smacks, Zulu and Fifie herring drifters and—in the 1830s—the first steamships.

Legend has it that the Pierhead organised a looting party of the most enterprising local wreckers and beachcombers when the first steamship came over the horizon. It was authoritatively put about, on the best intelligence, that the vessel was afire and there would be rich pickings when she went ashore, as she undoubtedly would. As so often, before and since, the Pierhead was confounded by a reality stranger than its wildest imaginings.

The Pierhead Skippers' musings have been interrupted on only two occasions. The first was in the year 1614, when the worthy burghers of Shetland's "ancient capital", the village of Scalloway on the Atlantic coast just seven miles west of Lerwick, came over the hill for a constitutional one Sabbath morning and found out what was going on: a straggle of warehouses, whorehouses and speakeasies had been hurriedly knocked together at the water's edge to provide diversion for the Dutch fishing fleet. Immorality abounded.

The Dutch had taken to basing themselves in Bressay Sound each year for the three months after midsummer when the Shetland herring were fat enough to be worth eating. The Bressay folk, enterprising capitalists that they were, had risen to the commercial occasions which this presented. A few like-minded locals on the "Mainland" shore of Bressay Sound had followed their example of "villainie, fornicatioun and adultrie", as a Scalloway court edict in 1614 put it. "Mainland", by the way, is the slightly confusing name given to the largest of the Shetland islands, although it is divided into many districts and peninsulas, each with its own distinctive name. The mainland, without a capital 'M', refers to the mainland of Britain.

To return to the 17th century Scallowegians: they were understandably vexed; if there were any spare guilders around, they were supposed to be directed to Scalloway via the traditional trading site at the Hollanders' Knowe—a prominent knoll in the ridge of hills half way between the North Sea and Atlantic coasts of Shetland. The upstart shanty town on the sewage-fouled shores of Lerwick, which in Old Norse means 'muddy bay', was put to the torch. Dire penalties were threatened to anyone who fancied setting it up again, on either side of the sound.

Fortunately, the Scalloway folk were so busy with their own moneymaking, villainy, fornication and adulteries over the next few years that they did not at first notice when the immoral little township on the backside of Shetland crept back in the night. Once

The ancient capital: Scalloway Castle was the seat of Scots power in Shetland during the 17th century. Here indignant grandees twice signed orders for the upstart Lerwick to be burned to the ground.
Photo: Malcolm Younger

more the sordid but highly profitable gin bothies and beerstalls were set up on the foreshore, in the common land of an otherwise entirely respectable Shetland hamlet called Sound. A further burning of Lerwick was ordered in 1625, but after that the high-minded Scallowegians (from whom the word "Scallywag" is supposed to derive, or so the Lerwegians say) appear to have given up the unequal struggle and turned their backs on the vice around Bressay Sound.

The bothies became a hamlet in their own right. After a hundred years or so the hamlet became a village. By the 1880s it was a proper little town, complete with such marks of civilisation as a Masonic lodge and several hundred paupers. All this time, boy babies were being born and raised in the art of Pierhead Skippering. By the time the oil industry arrived in Lerwick there were over 6,000 inhabitants, half of them self-opinionated male pundits on nautical affairs, occupying their accustomed places at the bars of the waterfront honky tonks.

These, then, were the kind of people who, with rolled up copies of the magazine *Sea Breezes* in their dungaree pockets and roll-up cigarettes in their mouths, beheld the oil companies' seismic exploration ships steaming into the port of Lerwick in the summer of 1969. To be sure, there had been a few of them calling now and again since 1965, but in that summer they became regular visitors.

The Pierhead's general opinion was that they were as ugly as sin and wasting their time. The first observation was correct. The ships, some of them converted trawlers, others former coasters, were anything but pretty. Even if they had once possessed handsome lines, they had long since been uglified, as various pieces of ironmongery were bolted and welded aboard them, with more thought to practicality and economy than traditional nautical aesthetics.

Their highly unlikely job was to tow cables around the sea, some of them two miles long. Inside the cables were microphones which picked up signals reflected from the seabed. The signals came from small depth charges set off by the survey ships at regular intervals. The echoes, logged on computers which might well be housed in ugly portacabins lashed to the deck, were analysed to give the petroleum geologists some idea of what the rocks were like beneath the seabed up to 500 feet below the surface of the North Sea. From this information they could work out whether it was worth drilling for oil and gas.

The ships were, in a manner of speaking, fishing for oil. In 1969 few thought they would find any. Their quest became the subject of ribald sarcasm among the worthies who frequented such longshore hostelries as the Queens Hotel and the Thule Bar. "Only oil they'll find oot yonder's fish oil, me boys!" was a typical comment of the times.

Little did the Pierhead Skippers know that within five years they would be jostling for space at their favourite bars, or that in the first decade of oil they would be elbowed out altogether by the boisterous crowds of roustabouts, scaffolder "bears", drifters and swaggering buckos of all descriptions, all lured by the black gold and spending their share of it in what were to become some of the most unpleasant bars in the northern hemisphere.

Old Lerwick hands lament more than anything the effect of oil on the pubs. The rutted roads, the housing shortages, the 12-hour shifts making beds for incontinent Glaswegian construction workers in the camps up at Sullom Voe—all that could be borne, but to see the destruction of the pub where you learned to fall over, this was a terrible thing.

A Shetland "foy" (party): some of the islands' musicians gather for a Saturday lunchtime session at The Lounge Bar in Lerwick. "Peerie Willie" Johnson, the legendary guitar player, is front right.
Photo: J. Wills Archive (photographer unknown)

Quiet, contemplative pubs of polished wood, with chrome bar fittings, brewers' mirrors, and framed cartoons of local worthies, became either spit-and-formica hellholes, loud with the din of fruit machines and other people's choice of tenth rate junk music, or gentrified, tarted-up "executive" saloon bars indistinguishable from their hideous counterparts the world over. In few pubs were things kept under control. The Lounge Bar in Mounthooly Street was one of them, the nursery of some of Shetland's legendary folk musicians, such as Aly Bain and Peerie Willie Johnson, and the university of several musical and gossiping generations of Lerwegians. Let us salute now the efforts of Hughie Robertson and his staff at the Lounge where the tide of oil rose and fell but the bar retained its essentially local and familiar character.

This may sound trivial and flippant. It is not. Changes to the bars and who was in them mattered a great deal to the islanders. They were a barometer of what was happening, even if you couldn't be bothered to follow the Byzantine intrigues of oilmen and councillors, reported at such length in the local and national press.

It is hard to keep a secret from the Pierhead. But the oilmen managed it. The discovery, by Shell, of the giant Brent oilfield in 1971, 100 miles northeast of Lerwick, was kept dark for a year. Commercial confidentiality, of course, but the long secret meant that the people of Shetland, who were to be most affected by the oilfield, had less time than they might have had to prepare themselves.

A year after the public announcement, in August 1972, that the Brent oil would be piped ashore to Sullom Voe, 25 miles north of Lerwick, Shell were still talking of a "£20 million oil terminal" to handle 300,000 barrels a day—a quarter of what Sullom Voe would eventually handle at peak production. Shell thought that there would be a Shell terminal and anyone else who found oil would build their own. They were wrong, but the sum of £20 million caused a sensation in a community where the local council spent nothing like £20 million in a whole year. Had the council suspected that Sullom Voe would eventually cost over £1,300 million, there would have been panic. As it was, there was consternation in the council chamber.

In a prophetic editorial comment on 11th August 1972, the week that *The Shetland Times* carried the front page headline "The Oil Gushes", the editor, Basil Wishart, had this to say:

> The first hard facts about the extraction of North Sea oil as it affects Shetland were revealed this week and it is now apparent that the social and economic effects will be quite as great as we forecast

> earlier this year. It should be borne in mind that the Brent field, as Shell-Esso have named the big find, is only one source of oil in the Shetland region. If it alone is allowed to inundate us with four hundred workers plus their families, it can easily be understood that the overall increase in population in this decade can be assessed as at least ten times that figure.
>
> It is almost pathetic that the Government, pinning its hopes on oil for the economic salvation of Scotland, seems to have left it to a relatively impoverished local authority to wrestle with the enormous infrastructural problems the oil boom has created. In our view there should now exist a department of the Scottish Office equipped to assist the local authorities in their planning, and a considerable fund from which to finance the housing, schools, roadworks and other developments an influx of population entails.

Four hundred construction workers? By 1980 it was to be 7,000. And the resident population of Shetland, after the builders had been and gone, was to rise not by 4,000 but by 7,000. The same issue of the newspaper carried a report on unemployment which showed that there were only 303 people registered locally as looking for work. No wonder that many islanders felt that they neither wanted nor needed the shock of oil. But matters were already out of their hands.

Shell's handling of the Brent discovery set the pattern for oil company public relations in Shetland: release information only when it suits you; play down the impact of the impending development; make reassuring promises about full consultation and co-operation with the natives; and send smooth-talking PR men round the village halls with slide shows and videos which flatter local culture and assure the audience that all will be well. Hornswoggling had begun in earnest.

The £1.3 billion they were to spend at Sullom Voe was recouped by the proceeds of a couple of years' oil production. Offshore, even vaster sums were to be spent as the industry, egged on by governments hungry for Arab-free fuel and huge tax revenues, threw money at its greatest technical challenge—extracting oil from thousands of feet below the seabed, using huge concrete and steel platforms which stood in wild seas up to 600 feet deep. The platforms had to be built, towed out, fixed to the seabed, commissioned, hooked up to hundreds of miles of subsea pipelines—and then serviced by helicopters, supply ships and standby safety vessels.

Shetland did not get a very large proportion of this "oil-related" business—Aberdeen, Peterhead and Dundee cornered the market early on—but there were still enough supply ships, helicopters, pipelaying barges and oilmen looking for beds to swamp the local labour market. An economy that was already overheating, because of the initial construction work at Sullom Voe, was to turn white-hot in the second half of the 1970s. Fortunes, large and small, were made. Sometimes the cash was drunk, but more often it was ploughed back into the Shetland economy. As the eighties ended, the islands were still awash with the money that oil had brought—not just to the local council (which could have had even more if it had played its cards better), but to dozens of commercial undertakings who sold goods and services to the oil industry and its hangers-on. Lerwick Harbour Trust, founded by those canny Lerwick burghers in the 19th century, was in there with the best of them from the start.

Urged on by their harbour clerk, the diminutive but dynamic Arthur Laurenson who had succeeded his father and namesake in the harbour offices, the trustees took dramatic steps as soon as the first whispers of huge oil developments were heard: they quickly cornered the local market in the valuable development land along the Lerwick waterfront.

Along the Sullom Voe waterfront things turned out very differently. Land speculators, backed by a consortium of Edinburgh merchant bankers, moved in and left the islands council with an eventual bill of £11 million for the development land which was to be compulsorily purchased by the council—after bitter battles in the council chamber and the Houses of Parliament, 800 miles away in London.

Lerwick Harbour Trust was composed in 1972, as now, of councillors with an interest in marine affairs as well as representatives of fishermen and local shipowners, and businessmen whose trade depended, to a greater or lesser degree, upon the success of Shetland's biggest commercial port. The trust's powers, as a public body established by Act of Parliament, and its astonishing commercial acumen are perhaps resented and envied by the citizens of Scalloway. Scalloway's harbour trustees were first in the game as an oilfield supply base, despite the minor detail of being on the wrong side of the islands for the East Shetland Basin oilfields. But by 1980 they had resigned their powers to Shetland Islands Council in return for promises of port developments which have still not all come to pass.

Meanwhile, in Lerwick, there were soon hundreds of shore jobs in the supply bases, pipe yards and warehouses backing up the extraordinary goings-on out there over the eastern horizon. Arthur Laurenson's grand coup was to buy up the derelict Gremista Farm, fringing the north-west shores of Bressay Sound, as early as 1972.

From then on, nothing happened on the Lerwick waterfront without the trust's say-so or without the trust taking some revenue.

Within 10 years the trustees had helped to develop four major oilfield supply bases, a new roll-on/roll-off terminal for the Aberdeen to Lerwick ferryboats, a new fishmarket and fishing boat dock, two industrial estates and, in 1989, a huge floating dry-dock designed to allow Lerwick to cash in on the demand for repair and maintenance for fishing vessels, Sullom Voe tugs, small oilfield supply ships and coasters of up to 2,000 tonnes deadweight.

By the early 1980s, they were making so much money that they had to build a luxury hotel in Lerwick to soak up the profits which otherwise would have been heavily taxed. By the end of the eighties, the hotel was contributing a quarter of the trust's income of £4 million a year. They even brought the derelict farm at Gremista back to life, winning top prizes at the Aberdeen sheep sales in 1988.

A customer at last—Lerwick Harbour Trust's £5 million oil rig repair base at Dales Voe sat idle for four years during the oil price slump of the later eighties—until this Swedish accommodation rig called en route to a contract at Sullom Voe. The *Safe Lancia* housed hundreds of men who repaired the massive corrosion problem at the oil terminal in 1987–89.
Photo: Dennis Coutts

For good measure, the trust also developed Europe's most modern base for inspecting, repairing and maintaining ("IRM") oil drilling rigs. The base at Dales Voe, just north of the town, was opened by the Prince of Wales in the summer of 1986. It was not the trust's fault that the opening coincided with a temporary collapse in the price of oil and, with it, the collapse of demand for "working-over" drilling rigs. At the time of writing, Dales Voe has yet to inspect, repair or maintain a drilling rig. It is in mothballs. But even this is no loss to the trust: the islands council was persuaded to part with £1 million for a new road to the base; British and European government agencies chipped in with several more millions in grants and, in return for its own relatively small tax- deductible investment, the trust persuaded a new consortium, IRM Shetland Ltd, to sign a seven year deal which guaranteed enough cash to pay for the trusts's share of Dales Voe, even if it was never used. If no rig ever docks at Dales Voe, the trust will be left with a useful pier for very large vessels, plus a big warehouse, offices and a workshop which can be used for other purposes in the future.

Other Lerwick Harbour Trust deals were just as clever: BP and Shell paid rent for land for their offshore supply bases, even while they were disputing rent with the council for land at Sullom Voe for over 10 years; local and incoming firms took space on the trust's industrial sites; P&O Ferries rented land for a new cargo depot and truck repair base; and even the oilfield service base that failed, at the Point of Scatland, was turned into a fish processing plant, to take herring and mackerel from the purse-seiners and trawlers that had previously been forced to sell their catches at rock-bottom prices to Soviet bloc factory ships. And the biggest oilfield supply base of all, Norscot, later taken over by Ocean Inchcape Ltd, was up and running on harbour trust land at Greenhead as early as late 1973, financed by the Norwegian shipowner Fred Olsen.

The casualty of Arthur Laurenson's commercial acumen was the council's former policy of spreading industrial development around the islands as far as possible. In the 1960s councillors planned to encourage industrial development based on "holding points" scattered throughout Shetland. The policy was backed up by the Highlands and Islands Development Board set up by the Labour Government in 1965.

At first, oil promised to bring this idea to fruition. Thanks to oil money, there was indeed to be some economic development in the "country districts", far from Lerwick and Sullom Voe. But much of it was precarious and mainly a matter of supporting uneconomic businesses with various subsidies, long past the point where the accountants and the banks would have shut the factories down. By the end of the 1980s, Lerwick was even more dominant than it had been at the end of the 1960s, despite Sullom Voe and the big helicopter bases at Unst and Sumburgh, at the northern and southern extremities of the islands respectively.

The town grew and grew, until it began to run out of building land, school places for the children of the new suburbs, and even fresh water for the kitchen and factory faucets. In Lerwick the market triumphed in decisions about where the jobs would be, as it did around the overgrown village of Brae, next to Sullom Voe itself, and in Unst and Sumburgh. Only in the late eighties, as the new industry of fish farming began its own spectacular boom, did the "holding points" philosophy begin to make economic sense as well as social sense, but by then Shetland was a far more urbanised and centralised place than it had been 20 years before. It also had developed the social novelty, for Shetland, of a large and successful middle class of entrepreneurs, managers and professionals.

From Day One of oil, the Zetland County Council, which merged with Lerwick Town Council in 1975 to form Shetland Islands Council, aimed to control the spread of the new industry—to plan the housing, roads, schools, public halls, waterworks and new drains it would need, to share the wealth around and, above all, or so it seemed to some early critics, to protect employers in the traditional industries of fish processing, knitwear and tourism from losing their labour force to the big money at Sullom Voe.

What happened was not quite like that; the oil companies did agree that they would all share the one oil and gas terminal at Sullom Voe—but only because it suited them, politically and logistically to do so; most of the new, planned settlements for oil were a social failure—ticky-tacky little boxes on bleak hillsides without trees to screen their designer-ugliness from the sodden blasts of a Shetland winter. Moreover, much of Shetland went through a chaotic decade of sordid trailer parks, shanty-town airports, single-track roads pounded into mud, power cuts, water shortages and social friction—not least in those overwhelmed pubs. Few architects or council planners lived in any of the four grossly expanded villages of Brae, Voe, Firth and Mossbank, which their profession, sanctioned by councillors, was to transform around Sullom Voe. None of these villages became large enough to sustain the range of services which even a small town in Scotland is expected to provide. They became characterless little dormitories, neither crofting townships nor thriving new settlements. With hindsight, a single new town at Brae would have saved a lot of bother. The option was considered but, alas, rejected.

Twelve years after first oil flowed, the islands are only now getting back to normal. It is not the old normality. Shetland has been immensely enriched by the oilmen and the money with which they were persuaded to part, but socially it has been torn apart and unevenly stitched together again. Without the efforts of the council, however, the damage would have been permanent and the wealth would have been concentrated in a very few hands, with little long-term social or economic benefit to the general population. That is the Shetland councillors' greatest achievement and the yardstick against which all their myriad follies, conceits and financial and political stumbles must be measured.

Back in 1972 and '73, we were solemnly assured that this development would be different. Unlike the shambles which had accompanied the oil industry in other remote oil provinces of the world, this time it would be planned, and planned in good time. All the on-shore back-up needed would be programmed so that the houses, roads, schools, airports, power stations, electricity lines, water pipes and drains would be there as soon as they were needed.

We were even told, in July 1974, that the ratepayers (local property tax payers) of the islands would not bear any of the cost of oil-related developments. The speaker was Ian Robertson Clark, the last county clerk of the old Zetland County Council and first chief executive of the new Shetland Islands Council which replaced it. He was reassuring councillors in Lerwick Town Hall as they signed the first of their troublesome agreements with the industry. It was a pledge which was to haunt him, his political bosses and their successors for a decade. What happened was that Shetland's public services ran up a massive debt to provide in five to 10 years what would have taken three decades at normal rates of spending. The interest had to be paid and the debt repaid. That bill fell upon the ratepayers; not just on BP, the biggest ratepayer, but on every little house and apartment in the islands. Rates were indeed abolished, on 31st March 1989, but only to be replaced by an even more hated tax, the "community charge" or poll tax, imposed by Westminster Government against the wishes of almost everyone in Scotland and in Shetland.

Yes, planning would cure all the problems of rapid development. That is what they said, and in 1975 the council set up a planning department, something not found necessary before oil. Well-meaning councillors, planners, civil servants, visiting sociologists and plausible oilmen told us not to worry. They probably even believed it themselves.

Young couples trying to bring up a toddler and a baby in a damp trailer park during a wet, windy and dark Shetland winter are not my idea of a planned development. And there were hundreds of them. These were not the "bears", the travelling men from Ireland and Central Scotland who follow the construction industry around the world—they were housed in purpose-built luxury camps and on accommodation ships where they wanted for little. They had entertainment laid on, plenty of rich food, beer on tap and warm and comfortable beds to collapse into, each man to his own room, usually. But young local residents, working the same 12-hour shifts for seven days a week, found that even their unheard-of wages were not enough to buy somewhere decent to live—because there was nowhere to buy. The boom moved too fast for the planners.

If Lerwick Harbour Trust was quick to see the possibilities of oil, so were the land speculators whom I mentioned briefly earlier. They found that much of the spadework had already been done for them by the council.

With commendable promptness, Zetland County Council had commissioned consultants in 1972 to work out where the oil terminal should go, where the pipelines should be brought ashore, and where the rest of the new developments might best be sited. These were the first paid advisers in what was to become a multi-million pound industry in its own right—telling diffident Shetland councillors what they ought to do.

It was not that Shetland's councillors were necessarily any more witless than their counterparts elsewhere in Britain—some of them were far from that—but they were used to dealing with the small beer of minor roadworks, bridge repairs, broken sewers, refuse collection and school meals. Their largest contracts had been valued in tens of thousands of pounds. Now they were dealing with hundreds of millions. It was all too much to handle on their own.

In the late summer of 1972, the council's first consultants, Transport Research Ltd of Glasgow, did a thorough job in a very short time. They discovered what the Lerwick Pierhead Skippers would have told them, had they been asked—that although Shetland was well endowed with dozens of natural and man-made harbours, only three were big enough for this job. Basta Voe, on the north east coast of the island of Yell, had been an Admiralty anchorage during the First World War—the base for a squadron of armed merchant cruisers. But it was altogether too remote, and a bridge linking Yell to the rest of Shetland was impracticable. For 16 years, however, it was to be the centre of a minor speculation of its own, as a local company tried and repeatedly failed to attract an oil terminal, ship-repair companies, oilfield service bases and other oil-related projects. In the end it became a centre for salmon farming.

Then there was Swarbacks Minn, a large, deep inlet with many tributary voes (narrow bays) and firths (wide bays); it too had been used by the Royal Navy, and it was handy for roads, electricity and water supplies. The problems were that its entrance was from the open Atlantic and, therefore, likely to be tricky in bad weather. Furthermore, most of the available flat land surrounding it was, by Shetland standards, in fairly active agricultural use and quite well populated. If crofters had to be shifted and dispossessed there would be big trouble, even with generous compensation, in a place where people's folklore memories of the 19th century landlords' evictions were, to say the least, vivid. Besides, Swarbacks Minn was on the other side of Shetland from the oilfields. That would mean longer pipelines and more disruption on shore.

That left the biggest and most sheltered deep inlet of all—Sullom Voe, on the other side of the isthmus occupied by the then sleepy little village of Brae. Sullom Voe was surrounded by undulating low hills, completely sheltered from all directions. It boasted a splendidly deep entrance and, best of all, was overlooked by the headland of Calback Ness, where the crofters had given up the struggle and left, long before the arrival of the electricity, tarred roads, water schemes and council housing estates which had saved many another crofting township from extinction. Besides, the long approach to Sullom Voe, down the broad channel of Yell Sound, was also sheltered in most weather except for the relatively rare gales from the north and north-east; a tanker had only to get in past the dramatic Ramna Stacks rocks which guarded the northern entrance to the sound. The passage was studded with many small islands and reefs but there was a good, fairly straight and comfortably deep channel for the largest ships afloat. And, once the few hundred sheep on Calback Ness had been cleared off, there would be plenty of room for the largest oil and gas terminal in Europe. There would be no need for new "clearances" of crofter Davids for the Goliaths of oil, only a handsome windfall for the local graziers and some big-spending companions for the handful of folk who still lived in the nearest settlement, Graven, a bleak spot surrounded by even bleaker ruins left over from the wartime flying boat base.

A further bonus was that the RAF's old airfield, built for Spitfire patrols at Scatsta on the south side of Sullom Voe, could be brought back into use. That would cost only a few million pounds. The

Tanker Approach to Sullom Voe

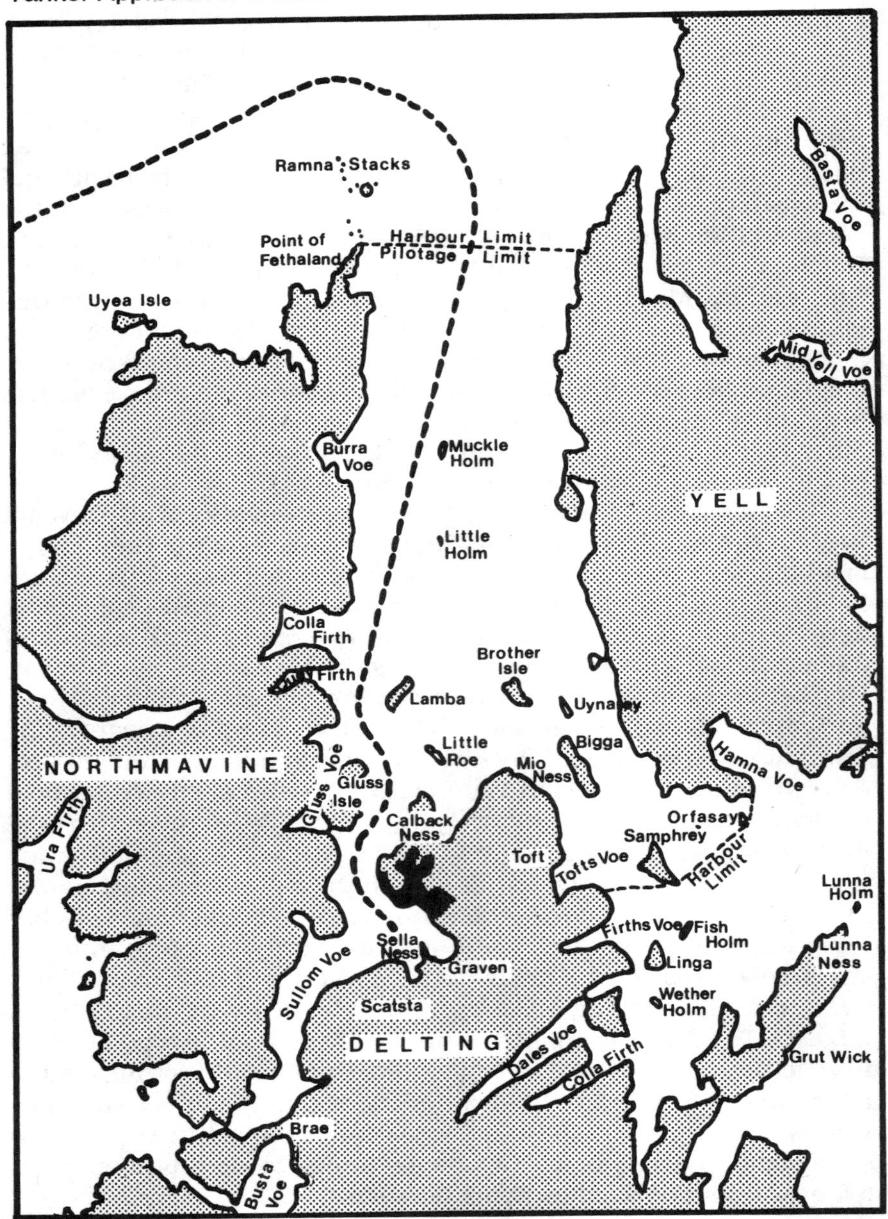

Cartography: Donald Battcock, from a map published in *Shetland's Oil Era* by Elizabeth Marshall, published by The Research and Development Department of Shetland Islands Council.

construction workers, to say nothing of the top oil company executives, could be flown to their own private airport, direct from mainland Britain. In 1978, four years after work at Sullom Voe began, the airport opened for business. So much for planning. Nearby, at Sella Ness, a low-lying peninsula provided a panoramic view of the voe—ideal for a port control operation and situated conveniently next to a small inlet where, for a few more million pounds, a harbour for tugs and pilot boats could be built.

Sullom Voe was ideal. And what the wise men of Transport Research Ltd came up with in outline was something very like what was eventually built. They dubbed the whole area "Nordport", which sounded suitably Scandinavian and explained the purpose of the project nicely. Councillors were mightily pleased when they accepted the consultants' recommendations and began serious talks with the oil industry about how all this might be brought into being, controlled and milked for the benefit of the locals. The oil industry in Shetland at that time meant Shell, the discoverers of the first big oil out at Brent, 110 miles over the horizon.

The late George W. Blance, the headmaster turned politician who led the Shetland council during the crucial deals giving the islands powers to tax and control the oil industry.
Photo: J. Wills Archive (photographer unknown)

There were three central planks in the council's policy, hammered into place by the then convener or leader of the Zetland County Council, a wise old freemason of covert socialist sympathies, George W. Blance, with a little help from his friends, to whom he was known as Dodie Wullie (Shetlandic for George William). Mr Blance, a retired Lerwick headmaster who had ruled generations of terrified pupils (some of whom were councillors by now) with a rod of iron, a voice of foghorn proportions and a well-concealed kindly nature, explained to Shell that the council would own the development land and lease it to the oil companies. Furthermore, the various partners in the offshore oilfields would have to share a single terminal—there would be no proliferation of individual, company-owned terminals, as at Milford Haven and Rotterdam. He made it clear that Shetland expected considerable financial benefits, not just for those who secured work building, running and servicing the terminal but also for the community at large, via their elected representatives including himself and his colleagues on Zetland County Council.

The oil companies cannot have believed what was happening. They were used to being welcomed with open arms or, in less populated parts of the world, doing exactly what they pleased, after placating the local tribal headmen. But it soon became clear to them that Mr Blance and his colleagues, including one Edward Thomason who was to become the council's leader 14 years later, were quite serious in their impudent demands. And they seemed to have the support of local public opinion and of the national news media, who had already scented a good David and Goliath story with undertones of Cinderella and the Fairy Godmother.

The council's "Nordport" scenario was being finalised when it was upstaged spectacularly, on 31st October 1972, by a rival private enterprise scheme, whose backers had the audacity to steal the name "Nordport".

Their local front man was Ian Caldwell who had come to Shetland as a hippy type in search of the good life and subsequently gone into business as a knitwear manufacturer. He announced himself as the "managing director" of the Nordport Company Ltd, funded by investment bankers from Charlotte Square, Edinburgh, who promised £20 million for the project. He was to be the first, and last, managing director. His proposals created uproar and immediately split the veneer of public support behind the council's scheme for Sullom Voe. Mr Caldwell, a giant of a man, well over six feet tall, contended that there was no need for the council to be involved at all beyond the normal planning controls which all local authorities in Britain possess. He offered the oil companies a "turnkey" project which would provide them with the land and services they needed to build a terminal, bring the oil ashore in subsea pipelines, store it in tanks, take the gas out of it and then ship out the oil and gas to market. As he pointed out, that was what had been done in most of the world's other offshore and onshore oil provinces to date. He had the backing of some of Scotland's biggest investment companies, notably investment managers Ivory and Syme and the merchant bankers Noble Grossart—both with their headquarters in Edinburgh's prestigious Charlotte Square, which is to Scotland what the City is to London or Wall Street is to New York.

Caldwell held what looked like a trump card: he had got his hands on the land, as councillors soon discovered to their utter dismay. Over the previous few months, Nordport and its proxies, using Charlotte Square money, had been flitting through the Sullom Voe townships, buying up land at unheard-of prices and paying for lucrative "options to purchase" from crofters and farmers who had never seen so much money in their lives.

The previous Sullom Voe boom, that occurred when the RAF arrived in 1939 to build the Spitfire runway and a flying boat base, had come and gone in six years, leaving them with a collection of mouldering Nissen huts which now did service as cattle byres, hay barns and implement sheds. The oil boom looked considerably more durable, and the locals were not slow to sign on the dotted lines. As one farmer said in the Orkney island of Flotta, chosen by Armand Hammer's Occidental Petroleum as the site of a smaller oil terminal at about the same time: "You can't stand in the way of progress, can you?" No indeed, not when you are offered £250,000 in cash to stand out of its way, as he had been.

The Scottish press came north to investigate, notably Chris Baur, then a financial correspondent with *The Scotsman* newspaper in Edinburgh and, many years later, the paper's editor for a while. I had the pleasure of helping him with his research, which resulted in the publication of Nordport's "family tree". This showed that the whole operation was a front for a get-rich-quick coup by the Edinburgh merchant bankers, using a bewildering network of holding companies and financial dead-letter boxes.

Caldwell and his backers affected hurt surprise at the revelations in *The Scotsman*. They argued that there was nothing wrong in coming forward promptly with the seed money for the biggest single industrial development ever seen in Scotland. On the contrary, they appeared to expect a pat on the back for doing their job—and doing it rather well.

Chris Baur's articles were to turn the tide against the speculators. PR men for BP, Shell and Conoco were quick to sense the way the political currents were running in Shetland and, by January 1973, when a liaison committee was set up between the council and the three oil giants, they were publicly expressing their support for a measure of local authority control. At the end of that month, the council hired more consultants—this time a firm called Livesey and Henderson, at what then seemed the enormous cost of £70,000—to carry out a detailed survey of the oil terminal site, followed in May by the appointment of planning consultants Llewellyn Davies to prepare Shetland's first county plan. Caldwell became nervous. By the beginning of February he was talking about co-operating with the council but this did not prevent them from turning down his application for outline planning permission at Sullom Voe on 16th February.

Word of the big events in Shetland was not long in reaching the City of London. By 23rd April the merchant bankers and investment managers N.M. Rothschilds had approached the county council with an offer of help. Their expensive financial expertise was accepted and, over the next 16 years, they were to be handsomely rewarded for investing the large sums of money which came the council's way.

Caldwell and his friends were to be defeated politically, if not financially, but not before some electoral blood-letting on a scale not seen in Shetland since the emancipation struggles which had freed the crofters from near-serfdom 90 years earlier.

It was one thing for the council to make plans for the welfare of both the islanders and the oil industry, but quite another to put them into practice, faced with a *fait accompli* like the Nordport Company. How were they to do it? The answer came from Ian Clark, a man who came to be loved and loathed in Shetland, in roughly equal proportions. In 1972, few people outside Town Hall circles had heard of the devoutly religious Mr Clark, still in his thirties but prematurely white-haired. He had come to Shetland as county treasurer and had been promoted to county clerk (equivalent to chief executive or general manager) after his predecessor left under a cloud.

Sizing up the political chaos around him, he decided that Shetland would have to acquire special powers from Parliament if the feared free-for-all with the oil business was to be avoided. There were precedents for this. Many organisations, such as Lerwick Harbour Trust, for example, had been established by Act of Parliament in the late 19th century and had their constitutions amended at intervals thereafter by a procedure known as "promoting a provisional order". After scrutiny by Parliamentary committees, such provisional orders usually become fully-fledged orders—regulations passed, in theory, by the Queen in consultation with her Privy Councillors (senior politicians and lawyers), but, in practice, agreed to by Government ministers and civil servants. Many docks, waterways, airports and major municipal enterprises in Britain had thus been sanctioned and given authority to borrow large sums of money for developments, and to levy charges on those making use of the completed harbours, railways, waterworks, bridges, public markets and so on.

Clark's provisional order was, however, unusually wide in its scope. It permitted the council to license all works in certain specified areas, not confined to Sullom Voe but extending to lands and coastal waters in other parts of Shetland where some oil development appeared likely and where the need for public control was anticipated. The provisional order empowered the council to act as port authority for Sullom Voe and for much of the Shetland coast, a function not normally associated with local councils in Scotland.

It gave permission for the council to set up special funds, separate from the usual "rating" account which paid for sewers, roads, schools, etc. These new funds were to be used for "the benefit of the inhabitants of Shetland", a phrase that was to acquire a very wide, not to say confused, meaning as the years went on.

Ian R. Clark—the first chief executive of Shetland Islands Council and the man who allegedly "hornswoggled" the oil industry.
Photo: J. Wills Archive (photographer unknown)

That was not all. The order provided for the council to levy harbour charges on ships, to invest its cash in the stock market and in public companies, and, worst of all, in the eyes of Caldwell and his cronies, it gave very wide powers of compulsory purchase over the very lands which they had just paid for. In short, it gave the Zetland County Council and its successor, Shetland Islands Council, unprecedented powers which amounted to a very large slice of home rule.

At first it looked as if the provisional order would go through Parliament unopposed. There was widespread cross-party support for it and great sympathy in Westminster for Shetland's plight, tinged with admiration for the plucky stand of Ian Clark and George

Blance. By no means all the local Tories were in favour of it, but the initial noises from Edward Heath's Conservative government were, if not encouraging, at least not discouraging.

The Government's attitude was soon to change to outright hostility, thanks to the intervention of the Reverend Clem Robb, a Church of Scotland minister in the village of Brae, right next to Sullom Voe. Mr Robb, as stocky as his friend Mr Caldwell was enormous (but just as energetic), was a founder of the "Shetland Democratic Group" in March 1973. The group, which included 10 serving councillors, was in reality a bunch of proto-Thatcherite Tories backing Caldwell and Nordport's private enterprise solution to the Shetland problem.

The cause of the Shetland Democratic Group was "freedom" and its chosen enemies were those councillors who, in the group's eyes, were attempting to make themselves the new "lairds" of Shetland, which is, indeed, what the council later became when it bought up several Shetland landed estates.

Tories have rarely dared to fly their party colours at local elections in Shetland, preferring to call themselves "independents" or to stand in the name of various short-lived ratepayers' associations or single-issue protest groups, such as the Shetland Democratic Group. This is a pattern common to many other parts of the Highlands and Islands of Scotland, springing from an ignorant desire to keep politics out of politics and, in part, from the naive view that you can vote for a candidate's personal qualities without considering his or her political philosophy.

Clem Robb, as the new group's most famous spokesman (he had already been on TV to tell the world about the dastardly left-wing plot to do the crofters out of their heritage), went on to have a word in the ears of senior Tory figures at their Scottish party conference, including the Prime Minister, Edward Heath, it was said. He filled those ears with blood-curdling tales of socialist dictatorship run riot and innocent crofters trembling in their beds at the approach of the new municipal lairds. The crofters, through their union, entertained his fantasies for several months, doing considerable damage to the case for local public control; they were unable to see the oil company wood for the council trees.

Mr Robb was an excitable fellow, as were many of his supporters. Their chorus of alarm proved so effective that many backers of the council's private legislation, including Edward Thomason, then the member for Unst, lost their seats in the county council elections in May 1973. It was a hollow victory; supporters of the moves to give Shetland extra powers to deal with oil retained the majority on the council and the Shetland Democratic Group, including their new councillor, one Ian Caldwell, faded into political obscurity. Caldwell was victorious in his own constituency of Aithsting on the West Mainland of Shetland where his main factory was sited, but with his Democratic Group having failed in its bid for a majority on the council, he resigned from Nordport soon after the May elections. Both he and Robb subsequently left Shetland, mission unaccomplished, and few missed them. Their swansong was heard in the evidence to a House of Commons Select Committee whose Tory chairman used his casting vote to emasculate Shetland's private parliamentary legislation. This resulted in a long and eventually successful campaign by Jo Grimond MP and others to have its more potent parts restored and to see it become law in the spring of 1974.

Before Mr Robb's alarms could take effect, the provisional order had run into a more serious obstacle: the quaintly-named Ways and Means Committee of the House of Commons took a preliminary look at the provisional order and decided, in March 1973, that it raised issues of such importance and complexity that it could not proceed as a simple administrative order—it would have to go through the full procedure of a Parliamentary Bill, being debated three times and examined closely by a select committee of four MPs.

It is less well known than it should be that the British Government has no monopoly over the legislation presented to Parliament. In practice, by controlling the amount of time available for debate, the governing party rules the roost, but besides their bills, there are Private Members' Bills put forward by individual MPs. A Private Bill, which is what Shetland's Provisional Order had to become because the Ways and Means Committee had said so, is a different animal altogether. In theory, any corporate body, or even any private citizen, can also promote a new law in Parliament—called simply a Private Bill. In practice, the procedure is so expensive that only big companies and public authorities can afford it. Private bills were occasionally used, in days gone by, to deal with particularly troublesome divorce and inheritance wrangles involving the British upper class.

The reason that such bills are expensive is that to get one you have to hire the services of parliamentary agents, a branch of the legal profession who specialise in this business. To the British State, the idea of laws being framed in language which the ordinary citizen might understand is anathema. It would put the lawyers out of business, for a start, and there are a great many lawyers in Parliament and Government. The Shetland council quickly secured the services of a firm of agents by the name of Rees and Freeres, none of whose partners has been seen begging in the streets of London lately.

Even with this more cumbersome procedure, Mr Clark still hoped that the Zetland County Council Bill (ZCC Bill), as it had now become, would have a fairly unruffled, if expensive, passage through Westminster. But then, in April 1973, Mr Tam Dalyell, the eccentric but high-principled Labour MP from West Lothian, half-way between Shetland and London, stepped in. Mr Dalyell was and remains a fierce opponent of "devolution"—the buzz word in the seventies for attempts to delegate power from central to regional and local government. This may have had some bearing on what he did in the House of Commons when the Shetland bill came before the House. Or, knowing Mr Dalyell, it may not.

The Shetland Bill certainly raised some interesting ideas about increasing the powers of local government in general, even if it was mainly just about arming the Shetland council for the battle with the oilmen. Mr Dalyell, being a principled constitutionalist, thought that Shetland's proposed Act of Parliament raised issues of such importance that it should be fully debated and investigated by a select committee, rather than passed on the nod. So when it came time for all those MPs in favour to say "aye", Mr Dalyell loudly said "no". By all accounts he was the only MP to do so, although he said he supported the Bill and only said "no" to ensure a full debate. Because of Mr Dalyell's procedural objection, the Bill would have to go through all the hoops, starting with the first meeting of the House of Commons Select Committee in June 1973. It would take many months and meanwhile the land speculators were consolidating their hold on Sullom Voe and the oil companies were rapidly drawing up their own plans for development.

If Mr Dalyell thought he was doing Shetland a favour, he was wrong. Jo Grimond, now Lord Grimond, the veteran Liberal MP for Orkney and Shetland, was enraged, as were most Shetland councillors. Mr Grimond accused Mr Dalyell of obstructing the Bill for the purposes of sensationalism. The position of the Shetland Labour Party, which counted several councillors among its ranks, was confused, to say the least. I was their prospective Parliamentary candidate for Orkney and Shetland at the time and I can report that "enraged" is putting mildly what many of the party members in Shetland thought of Mr Dalyell's unsolicited intervention. Embarrassed is another word that springs to mind, to say nothing of homicidal. Alas, our fellow party member, Mr Dalyell, had not thought to consult his colleagues in the far north before chucking a spanner in the Parliamentary works. As so often before and since, he had gone off at a tangent, on his own. But, party solidarity being what it was, I was obliged to defend his actions, in a letter to the press which now makes embarrassing reading.

With hindsight, Tam Dalyell did not make all that much difference to the Sullom Voe timetable. For we had all over-estimated the oil companies' abilities. Like them, we believed that the oil terminal could be up and running in a year or two. A combination of North Sea weather, organisational foul-ups, millions of tonnes of wet peat on the Calback Ness site and technical teething troubles delayed the completion of the oilfield production platforms, the pipelines and the terminal itself. At the time, I was calling for a one year moratorium on oil developments, to give us all time to think and plan. Murphy's Law ensured that in fact we had five years before the first oil flowed in October 1978.

To be fair to Tam Dalyell, his extraordinary intervention ensured that when the Zetland County Council Act (ZCC Act) did reach the statute book, in the spring of 1974, it was a much more carefully thought out piece of work than the hastily assembled provisional order. It gave less offence to the various objectors, who all eventually withdrew from the fight, and it had greater Parliamentary authority than a mere order could ever have had.

The powers granted to the council remained substantially the same, but the area of land zoned for compulsory purchase was smaller—restricted to the terminal site and some other nearby land. Baltasound harbour in Unst was excluded, as was Basta Voe harbour in Yell, soothing fears of a council takeover in those areas. Interestingly enough, the owner of the only pier in Baltasound was to attempt to sell it to the council many years later, hoping that it could thereby qualify for expensive improvements which he could not afford himself. The Tory chairman of the Select Committee, Sir John Gilmour, at one point used his casting vote to delete the compulsory purchase clauses, while leaving in the powers of the council to control the tanker harbour, but his last-ditch obstruction was eventually over-ruled by the Government after lobbying by Ian Clark, Jo Grimond and the Labour Party.

Tam Dalyell was not the only Labour MP to cause the local party some embarrassment in those heady days. There was also Dr J. Dickson Mabon, who in 1979 was to defect to the new Social Democratic Party and to follow them into political oblivion 10 years later. Dr Mabon, along with Mr Dick Douglas MP and the right-wing anti-monarchist Labour MP Mr Willie Hamilton, paid a fact-finding visit to Shetland early in 1973. As the prospective candidate, it was my job to show them around. Among other places, I took them up to Sullom Voe in a Lerwick taxi.

Sullom Voe was still a calm backwater, beautiful to me and others for its birds and scenery, but no doubt rather bleak to these urban parliamentarians, shivering in a keen easterly drizzle. We drove down the old runway of the derelict RAF airfield at Scatsta, overlooking the site of the proposed oil terminal. When we reached the end of the runway we parked and Dr Mabon got out. He took a few steps into the wilderness, wrapped in a fur-collared overcoat and, if I remember rightly, sporting a fur hat. Taking a deep draw on his cigar, he gestured as expansively as his tiny frame would allow and uttered the unforgettable words: "What a splendid site for a major petro-chemical complex!"

Unfortunately, the taxi-driver heard this and the argument which subsequently developed between me and Dr Mabon. Being a local man, as most Shetland taxi-drivers were in those days, and not being of the Labour persuasion, he had the story all over town by nightfall.

My problem was that the policy of the local Labour Party, and of most other people in Shetland, was to restrict the scale of this fearsome industry as far as possible. We thought we might survive a huge oil and gas shipping terminal where tankers could collect the stuff and take it away to the refineries, but the last thing we thought we wanted was "a major petro-chemical complex". "Downstream" development, so beloved of generations of Labourite ministers, keen to encourage big business in areas of high unemployment, was neither needed nor wanted. When, in April 1973, a man called Warren Shear and his US company, Milford Argosy, proposed an oil refinery for Shetland, there was near-hysteria in the correspondence columns of the local press.

Poor Dr Mabon was only innocently expressing a widely held view in the party which was about to form a Government—all industry was good news because it meant more jobs, and the Government's job was to help it. That had been the thinking behind such state-sponsored ventures as the Linwood car manufacturing plant near Glasgow, the aluminium smelter built at Invergordon, near Inverness, in the 1960s, and the huge pulp mill set up at Fort William at about the same time. "Small is beautiful" was still a long way off and ecology was still something dabbled in by only a handful of eccentric geographers in the universities.

My argument with Dr Mabon was mirrored by the split in the Shetland Labour Party as the oil age arrived. Our debate in turn reflected the arguments among the rest of Shetland's inhabitants, few of whom voted Labour, then as now, although the party's standing has improved since I ceased to be its candidate. Alas, it was mainly a split between youth and experience. There were legendary Labour councillors such as the late Alex Morrison, Jimmy Paton and John Butler. Although they were my friends, they had difficulty in understanding my ecological perspective—for what was I, if not an eccentric academic geographer, at that stage of my career? They probably put it down to romanticism, induced by spending my childhood out of Shetland and by an unhealthy interest in bird-watching.

Alex and Jimmy had lived through the depressions of the thirties and the fifties in Shetland. They had been scarred by the unemployment and emigration of many of their childhood friends. They had welcomed the development of the Shetland economy in the sixties, when fish processing, knitwear factories, sheep-rearing and tourism had at last begun to support a reduced population in something like decency. But the wages were still low and the economy, although superficially healthier than it had ever been, was still fragile. It could easily collapse again because of capricious markets in the south, or be crippled by increases in sea freights and ferry fares. No-one foresaw then that, within 20 years, fishing would be in crisis because of gross over-exploitation of stocks, or that over-grazing would become a problem in Shetland's productive hills.

"Get the boys jobs!" was Alex's motto, and it served him well. A shopkeeper by trade, if not by inclination, he was delighted at the coming of oil. This was Shetland's chance at the big time, the islanders' revenge for decades of decay, depopulation and the petty tyrannies of small-town employers.

At the height of our private rows about the possible effects of oil on the Shetland environment, Alex once roared at me: "Wha's carin' aboot twatree [a few] birds? Get da boys jobs, dat's da ting! An' decent jobs, no like warkin' for some damned shopkeeper for bugger all!" Unlike some councillors with whom I have tangled, Alex did not bear grudges. Our profound disagreement about oil never clouded our friendship. But others were to condemn me when, some years later, I reported the fact that the council had fouled up the administration of its oil money reserve fund and put much of it at risk from the tax collectors. I was accused of "losing Shetland £20 million" and of disloyalty to the islands. If the emperor had no clothes, it was not my job to say so. This attitude, I am sorry to report, is still prevalent in some quarters in the late eighties, not so much among the freemasons, of whose role in Shetland more later, as among socialists, liberals and democrats who ought to know better. However, a wangle was agreed with the tax inspectors and the council was allowed to keep its cash, with spectacular results, which we shall examine in some detail.

I am running ahead of myself. Back in 1973, Nordport and its proxies still held the land. A key clause in the ZCC Bill gave powers for the council to acquire it by compulsion, as was eventually to happen. Nordport's financial heirs and successors, notably the merchant bank William Brandt, were to be paid some £3 million for it. What Ian Caldwell got for his efforts I cannot say and at this late stage I no longer care. By the time the council recouped the money from the oil industry, interest payments had taken the cost of acquiring the land to nearer £11 million at late 1988 prices. By that time such sums had become inconsequential, even in Shetland. The clever thing was that the oil industry paid for all of this. That was not a direct result of the 1974 ZCC Act but because of private agreements between the council and the oil companies, which we shall examine in some detail in Chapter Seven.

The Simple Life 2

Oil brought Shetland many things besides oil—more money, modern amenities, better transport, more publicity and, above all, more people. For the first time since over 20,000 unwilling troops had garrisoned the islands during the Second World War, it brought north thousands who did not want to be here. Many of these temporary residents did not like the place at all, did not intend to stay and were not slow in saying so. For the local custodians of our "way of life", this was a profound cultural shock.

Until 1972 most of our visitors and settlers came here because they were interested in the place. Those who fell in love with it, despite our weather, were eager to become part of the local community and lost no time in telling us all how wonderful we and our way of life were: the "characters" in the pubs; the quaint customs of the midwinter fire festival booze-up, Up Helly A' (and the midsummer booze-ups of the regattas and agricultural shows); the dramatic landscapes; the superb birdlife; the sturdy descendants of Vikings in their fishing boats; the traditional fiddle music; the long winter evenings of storytelling around the peat fire or dancing in the village hall; the seasonal rhythms of crofting; and even the weather, which at least was full of variety; all these were wonderful. Visitors and incomers said so constantly. A cosy conspiracy to romanticise us and our islands was woven around the central theme that here was a simple way of life, untainted by the evils of 20th century urban existence.

From time to time stray journalists, radio interviewers and even, marvel of marvels, television film crews landed to interview the natives and carried south a reassuring pastiche of island living, compounding the great lie at the heart of this arcadian libretto. "Characters" queued to be interviewed, "turns" were obligingly per-

Ready for a viking foray: Bressay ferry skipper Billy Clark in full pillaging kit on the morning of the island's Up Helly A' fire festival.
Photo: Dennis Coutts

formed on the fiddle for the cameras and microphones, the knitting needles clicked as the mawkish dialect poetry was recited—and local political worthies, with a small 'p', of course, pontificated about why Shetland was so very, very special.

John McGill, a Glaswegian teacher of English who spent many years at the Anderson High School in Lerwick, summed it up in a disturbingly sharp little poem:

> We need collectives: fury of fing
> ers knowing on needle or string fleet of fish
> ers sunsetsailing cot of crofters wish
> ing each to be the one the camera ling
> ers on for summingupshot creditlists
> and Magnusson's farewell. What if a muddle
> of mediamen be more than fair, the huddle
> of houses tighter than their brief insists?
>
> No matter, for their eye's a choosey eye
> that blinks at all the quids and quondams,
> takes in the fulmar's flight, but passes by
> the roadside clutter of cans and condoms.
> And for their ear the old wise tongue unleashes
> cataracts of current clichés.

Alas, it was true; by 1972 many Shetlanders lived either in the town of Lerwick or in council houses in the country districts. The old self-sufficient crofting life had died more than a generation before. Indeed, the "Golden Age" of crofting only existed for fewer than 30 years—between the emancipation of the crofters in 1885/86 and the outbreak of the 1914–18 war, which changed Shetland for ever, as it did almost everywhere else. By the 1930s most of Shetland was firmly in the thrall of the urban economy of Lerwick, Aberdeen and parts south, as Professor Andrew O'Dell had shown in 1933 in his pioneering study for the Le Play Institute, *The Urbanisation of the Shetland Islands*. But the myth-makers paid no heed to academic research, and so they continued to warble soothing and flattering scripts, from their concrete jungles in Lerwick and London, about the simple, self-reliant, noble savages of Shetland. It was what the outside world wanted to hear and, I need hardly tell you, the inhabitants lapped it up with enthusiasm. It was pleasing to be thought of as fine fellows, and charming and hardily Norse with it. Besides, it encouraged even more tourists to head north to wonder at the rustic simplicity—and to be quietly laughed at by the natives for their gullibility.

There was, and is, nothing simple about the Shetland way of life. I did once lead a simple life, when I lived in the city of Edinburgh: I went to work each day for eight hours or so, doing just one job which I knew quite well; the money I earned from this was my only source of income—nice and simple; I used it to purchase the housing, clothes, food and services which I needed; cash was all that was necessary to live in some comfort. I need have no contact with people unless I wished to; my circle of friends was established and I saw them by arrangement or, sometimes, when I bumped into them in the street. All very simple. Compare this with Shetland where, until recently, many people had to take several part-time jobs to make a living on top of what their crofts and fishing boats provided. Nowadays, the croft's produce supplements what is bought with wages, whereas it used to be the other way around. And fishing is very much a full-time occupation, although there is much casual labour on the new fish farms, which fits in well with running a croft. But it is still far from a simple way of life.

I thought it was simple until I went to live on that croft in Unst, back in 1973. To live there at all I had to have several sources of income, most of which involved daily contact with other crofters for whom I did occasional work, such as tarring roofs, helping with herding sheep and harvesting the hay and oats—work often done not for cash but in return for a leg of lamb, a bundle of spare fencing posts or a barrel of "buckshee" diesel fuel. Then there were the six men with whom I worked in the boat taking supplies and light-keepers out to the Muckle Flugga lighthouse. All this immediately involved me in a network of mutual obligation which no sociologist could ever hope to understand fully—I didn't understand it myself. For those who had been born and brought up in the community of north Unst it was even more complicated, for they carried the mental burden of generations of family feuds and neighbourly disputes, details of which had to be referred to instantly in their heads when considering every request for, or offer of, assistance. It worked, after a fashion, but it wasn't simple.

In addition to all these complications just to earn a living, I had to be my own carpenter, electrician, plumber, machine-mender, motor mechanic, slaughterman, roof-slater, glazier, house-painter, fencing contractor, boat repairer and peat cutter. North Unst was a Do-It-Yourself society. Even if I could have afforded it, many of these services were not available commercially in Unst or, if they were, the professionals were so busy that it might be weeks before they could come to the rescue, particularly in the summer, when a 16 hour working day divided between various jobs was by no means uncommon. Needless to say, my skills varied and never came near the astonishing performance of dozens of neighbouring crofters who could turn their hands to anything and make a workmanlike job of it or, often, a great deal better than that. So I had to call on their assistance.

Few would take money for their services and I soon learned not to make more than a ritual offer of cash. What I was expected to do in return was to lend a hand with the simpler tasks where my urban hands could be trusted—dipping their sheep, saving their corn from the autumn gales and babysitting when they had a social function to attend, although a refreshing feature of Unst social life was, and remains, the habit of taking even the smallest children along to weddings, concerts and dances in the local hall.

I soon discovered that I was poorly equipped to lead this "simple" way of life and was left with a wondering admiration for those who could handle all the social and practical demands made upon them by this curious and endlessly diverting island, a society half-way between the subsistence economy of the 18th century and the capitalist nexus of the late 20th century.

The stress which can result from failure to cope with all this is obvious, but so are the attractions of a society which immediately appears wildly romantic and inspiring to those who have only a brief acquaintance with it. And, when the sociologists and social anthropologists arrived in force in the early seventies, to study what made Shetland so special and, it was assumed, so vulnerable to Big Oil, they found that most Shetlanders were already well aware of the value of their heritage and becoming more so with each TV documentary about the place.

No wonder, then, that when Big Oil came we flew into a veritable frenzy about the threat to our way of life and redoubled efforts to tell the media of our ethnic treasury, of our uniquely admirable fix for the woes of this lamentable century. The unstated assumptions behind this cultural panic are revealing: the idea was that the sudden influx of large numbers of strangers must be destructive of local speech (Glaswegian speech, in particular, is thought of as "ugly" by some Shetlanders), and would be a threat to traditional music and do-it-yourself entertainment and an attack on "traditional" patterns of life in general. The "soothmoothers"—those who had come in to Shetland through the South Mouth of Lerwick Harbour—were about to descend like a plague of locusts.

What actually happened was so extraordinarily different from the perceived threat that it is hard, 17 years later, to take seriously the fears that were then expressed in the local press and to visiting TV cameras. What happened was that the oil age coincided with an unprecedented renaissance of Shetland's music, dialect writing, arts and crafts—and a big revival of interest in local history, folk culture and rural ways of life.

Many of the incomers were fascinated by traditional Shetland fiddle-playing, for example; soothmoothers took leading roles in local musical societies and in setting up the islands' hugely successful annual folk festival. They also made their contribution, loud-mouthed and opinionated as they might appear, to local politics—in new community associations, tenants' groups, hall committees and parish councils.

Interest in traditional music had survived the arrival of television in 1964, but only because of the efforts of the Lerwick-based musicologist and fiddle teacher Dr Tom Anderson and a few, remarkably few, like-minded souls who devoted themselves to preserving the old tunes and styles of playing and, crucially, to teaching younger Shetlanders how to do it. Pop, country-and-western and Scottish fiddle and accordion bands were far more popular among Shetlanders in the 1960s and early 1970s than traditional music.

It was Dr Anderson, through his research, who demonstrated that Shetland fiddle music, far from being a quaint, fossilised art form developed in isolation, was in fact an extraordinary amalgam of music from Scandinavia, Scotland, Ireland and further afield. With the Shetland Folk Festival, and dozens of other organised and spontaneous musical events, Shetland became a musical crossroads in the North Atlantic, rather than an insular backwater. Without the contributions of soothmoothers, as performers, audience and backroom organisers, Shetland's musical renaissance would have been much the poorer.

To the surprise of many, the new settlers were not, in general, hostile towards the distinctive Shetland dialect. They found it curious and, at first, difficult to understand, but also very interesting, even if their interest was limited to finding out the meanings of the islands' thousands of Norwegian-derived place-names. Sales of dialect poetry improved, along with the literary standard of some of it; the *New Shetlander*, a quarterly literary, political and historical magazine, found more readers; a new monthly magazine, *Shetland Life*, which specialised in dialect writing and local historical miscellanies, counted soothmoothers as well as locals among its subscribers; and publication of new books on every aspect of Shetland, from geology to hand-knitting, from ornithology to autobiography, reached unheard-of figures, with several worthwhile new books each year about Shetland, rather than one or two a decade, as in the pre-oil days.

Exhibitions of work by local artists also increased, with painters and illustrators queuing for a chance to display their works in the cramped quarters of the Shetland Museum Gallery in Lerwick and frequent small exhibitions elsewhere in the islands. Hotels and pubs showed local work on their walls and the demand from the general public for Shetland paintings, by no means all of them worth the canvas they were painted on, it must be admitted, increased beyond the wildest dreams of the daubers. Craft fairs in rural halls became a regular feature of island life and provided a showcase for the talents of local and incomer craftworkers alike—as well as a meeting place for the exchange of ideas.

In short, Shetland discovered that it had a wealth of artistic talent and began to value its own musicians, artists, woodworkers, actors, poets and writers just as highly as the metropolitan varieties imported via the TV, pop radio and the missionary efforts of the Scottish Arts Council's travelling theatre shows and exhibitions. The islands became culturally lively as never before—and much more self-confident. The answer to the old question: "But what do you do in the long winter evenings?" was now: "Well, we try to think up excuses for not going out to yet another concert, play, film, exhibition, festival or dance—anything for a quiet night at home!"

All this activity coincided with the oil era—but how far was it a direct result of the changes that oil wrought in Shetland? The answer is in four parts: firstly, the oil-related settlers turned out not to be invading barbarians but, more often, an admiring and astonished new audience who sometimes valued Shetland's culture more than the Shetlanders themselves; secondly, the perceived threat re-awakened Shetlanders' interest in their own culture and spurred them to join in the efforts to preserve what was thought to be threatened; thirdly, the oil cash enabled Shetland Islands Council and its various trusts to pour unprecedented funds into artistic activities—money for tuition in traditional fiddle playing was no longer a problem and the new Shetland Arts Trust became an important local patron of the arts; and fourthly, the general increase in prosperity, largely based on oil-related jobs, meant that the local public had spare cash to buy the concert tickets, records, tapes, books, paintings and craftwork which flooded onto the market.

One of the victims of Shetland's cultural revival was the small-minded belief that what came in the south mouth of Lerwick Harbour was either more fashionable and desirable than the local product or, a belief sometimes held simultaneously, a dangerous alien threat to everything locals held most dear, or were supposed to hold most dear if they were loyal Shetlanders. The money for painters, musicians and craftworkers based in Shetland was good but even better was the mingling of styles, materials and ideas in events such as the folk festival, the craft fairs and art exhibitions—to say nothing of the burgeoning talents revealed in displays put on by rural visitor centres/museums by locals and incomers in Unst, Yell, Northmavine and elsewhere.

It is hard to imagine all this happening without oil—without the influx of new people and the access to "petro-pounds". The benefits to Shetland outweighed the problems. One perceived problem was that in some school classes you could find a majority of pupils who were not native speakers of the dialect. This caused concern among many Shetland parents but my own experience, as the parent of two English-speaking teenagers, is that they feel obliged to adopt (and adapt!) local dialect words, rather than their classmates trimming their speech to English norms. Some of our "way of life" enthusiasts greatly under-estimated the dogged durability of the dialect; they also under-estimated the extent to which Lerwick speech, in particular, had absorbed southern usages in the decades before oil arrived.

The proof of the persistence of dialect or, at the least, of a pronounced accent, can be heard in the school playgrounds of Unst, where there has been a large population of incoming military families for nearly 40 years, and even in the suburbs of Lerwick. A tongue which had survived one and a half centuries of determined oppression by schoolteachers, many of them native Shetlanders, ordering their pupils not to "spik Shetland" had little to fear from oil. To be sure, the accents of London, Glasgow, Cardiff and even New York and Bergen may now be heard in our schools, but the pupils are none the worse for this touch of cosmopolitanism.

In these more enlightened days, few schoolteachers insist on being addressed in "proper English", although the conventional standard English is still taught, to satisfy the requirements of the examination boards. The exam results of Shetland pupils, not least in English exams, suggest that they are well able to cope with what amounts to bilingualism, and few have problems making themselves understood when they venture into the south.

Education was a major beneficiary of oil—not so much through the charitable trust as from the council's rates, or local property taxes, paid from Sullom Voe, which enabled the education service to be improved and extended in Shetland at a time when public education was suffering severe cutbacks elsewhere in Britain. Most districts in Shetland have new or greatly expanded schools, many of them next door to new leisure centres and community halls which astonish visitors who expect to find few modern amenities here. Inside the splendid new school buildings there is a wealth of equipment which would be the envy of many private schools in England, while the ratio of teachers to pupils is often better than the minimum laid down by a parsimonious Government.

The charitable trust cash is used for some educational bursaries but its main effect on youngsters is through lavish funding of sport and extra-curricular activities. Island teams regularly compete on the mainland and in other island groups as far away as Faroe, their travel is paid for by the trust; new sports halls and swimming pools are dotted around Shetland—three such centres, each costing over £1 million, have been built in Unst, Whalsay and Yell—islands with around a thousand people each—by the Shetland Recreational Trust, another "pensioner" of the oil money. The aim of the council back in 1972, to ensure that the benefits of oil money were spread widely through the community rather than accruing only to commercial interests and oil industry employees, has very largely been achieved.

Spectacular examples of the success of this policy are in public health and social work, where a miniature welfare state has been set up, making good the damage done by 10 years of Thatcherism and improving on the threadbare safety net which the state used to guarantee to all its citizens.

The most frequently-quoted example is the £200 annual Christmas bonus to every household where there is a resident pensioner or disabled person. But that is just for starters: pensioners living alone have home helps, specialist nursing at home when required, a telephone alarm system with bleepers to call for help from a continuously-staffed central switchboard, as well as a smoke detector in each house; in the country areas hundreds of pensioners' houses have been brought up to basic civilised standards with bathrooms, kitchens and lavatories paid for by the charitable trust; there are discretionary grants in cases of real need; disabled people have first-class day care centres, specially adapted Volvo cars, subsidised holidays and "respite care" centres so that relatives who look after them can also enjoy a holiday while their dependants are looked after for a week or two.

The loneliness, poverty and stress which were very much a part of the way of life for old or disabled folk in Shetland before oil have very largely been eliminated. Now they can look forward to a more decent life, with less worry and want, rather than a mere existence on the margins of society. Much warbling has been heard about the wonders of the old Shetland society where relatives in need were cared for by their extended families. But it meant enslavement and a stunted social life for the people, mostly women, who had to do all this unpaid caring.

Shetland's new welfare municipality is a big improvement, while retaining the best features of the old extended family. It also makes extensive used of those charitable citizens who work for nothing in the myriad voluntary groups in the islands.

Not all of this has been due simply to the oil money—much of it is the result of enlightened attitudes among councillors and hard, clever work by dedicated public officials who have managed to milk the Government's inadequate regulations for every penny they could squeeze out for the benefit of their disadvantaged clients. The system is a triumphant success and puts the British Government to shame. The philosophy behind it is to help the elderly, the chronically sick, the handicapped and the unfortunate to make as good a life as possible for themselves, while staying within their own communities. For example, although there is still a shortage of geriatric beds in the local hospitals (thanks in part to an appalling planning blunder by the local health board which is not controlled by the council but appointed by the Government), it would have been far worse but for the oil cash and the imaginative and compassionate uses to which it has been put. If you want private schooling for your children or private health care for your family, they are not available in Shetland: the public services are so good that there is no demand for commercial health care and education. It is no doubt very comforting for the Government, as well as politically convenient, to hear Shetland's councillors endlessly reciting that they are "a special case". They may be, but they should not be. What has happened in Shetland schools and in the Shetland health service and social work department has to happen nationally if Britain is to be re-civilised after the trauma of Thatcher's decade. After all, the Government has "oil money" as well—but has spent much of it on maintaining the unemployed on miserly incomes.

As a postscript to this chapter on the way of life in Shetland, and its lack of "simplicity", it is interesting to note the changing attitudes to litter. Shetlanders chuck things out, like everyone else, when they have finished with them. And these days they have almost North American quantities of junk to dispose of.

Traditionally, rubbish was thrown on the midden, or muck-heap, outside the croft door. It was part of the way of life and a very useful part too. The country folks who migrated to Lerwick in the 19th century took this habit along with them with the result that the town suffered serious public health problems and its city fathers, like many others in Britain at that time, were obliged to institute some basic measures of sanitation.

Most of what was chucked out used to rot—a key point when those worthy Victorian burghers of Lerwick hired the first scav- engers, "scaffies", and started thinking seriously about piped water and drains. Well into the 20th century most trash was still rottable and therefore of benefit if it rotted in the right place and was thus returned to Mother Nature. What did not rot would rust in time or, if it was glass, would be ground into rounded, harmless lumps of silica on the beaches.

Then came the worldwide plastic explosion in the 1960s. As late as 1970 the beaches of Noss, the nature reserve where I was warden in that memorable summer, were largely free of plastic. And clean driftwood, free from tar and oil, was plentiful; seaweed could be laid up to rot down for manure without spreading thousands of tiny pieces of plastic foam on your garden. But all that was to change in Shetland, as elsewhere.

Winds and tides brought Shetland waste plastic and oily tarballs from all over the North Sea, and even from North America, as the international shock waves of the plastics explosion ended up in the sea and added to the islands' home-grown waste problem. Then came the oil industry with its profligate junk, which brought Shet- land beaches even more plastic sheets, plastic drums and cans, plastic pipes and a bewildering assortment of jagged metal objects, some of them 30 feet long. And the oil slicks, the biggest since the Second World War—not at first the oil industry's oil from Sullom Voe but oil spilled from the increased volume of passing shipping and, among it, the tarballs which are what oil eventually becomes in the sea. Millions of them.

The spectacular accumulation of trash on land rose in concert with the marine pollution. Sometimes it seemed that Shetland crofters were vying with each other to demonstrate their new prosperity, by surrounding their houses with more junk cars and agricultural machinery than their neighbours. If you can afford to throw away that much, they may have thought, then you must be doing well. Tourists began to write rude letters to the papers about it. Most people laughed; a midden was still a midden, even if it was made of plastic, steel and aluminium. But it kept on growing and could never be spread on the ploughed fields although much of it contaminated the pastures. Oil meant money and money meant junk.

The oil industry, to do it justice, eventually cleared up most of its trash in and around Sullom Voe, although it left a lot more lying on the seabed, infuriating the fishermen who complained, and still complain, of miserly and tardy compensation for torn nets and lost fishing grounds. But the people of Shetland, and especially their elected council, did little to clear up the nest that they had fouled themselves.

The trouble was—where could you put all this stuff? Traditionally, old tractors had been dumped over the cliffs or, worse, over the low shores which make up much of the Shetland coastline. Many of them are still there, 30 or 40 years on. At one point the council even spent large sums of money making roads to coastal dumps, with neat concrete kerbs to stop you tipping your trailer over the edge, together with the garbage. Bressay was a particularly severe case of this municipal vandalism, with results that can be seen to this day at the scenic and historic Voe of Culbinsbrough.

There was no scrap merchant in Shetland and the traditional dumps in disused quarries were becoming eyesores that even Shetland Islands Council, famous for its eyesores, could no longer ignore. And then an enterprising firm arrived from Durham in the north of England and started up in business, feasting on Shetland's mountains of scrap metal. But even they could do no more than dent the vast piles of trash, still less tackle the thousands of scattered pockets of rusty junk festooning crofts all over the islands.

The answer was the Shetland Amenity Trust which was set up by the council with oil money, and cashed in on a Government training scheme which, like most Thatcher Government training schemes, was a thinly-veiled source of cheap, otherwise unemployed labour. An enterprising Canadian, Rick Nickerson, was appointed and set to work. Refuse skips for bulky waste were set down in country parishes and promptly filled to overflowing, long before they were due for collection by the trust; some of the contents of the skips were re-cycled by local beachcombers, who hated to see useful things like old baby buggy wheels thrown away. The Do-It-Yourself society was not dead yet. Trucks were sent out to make a start on salvaging the croft car wrecks, many of which were jealously guarded by their indignant owners, who regarded them as assets, should they ever have need of a spare windshield, dynamo or headlamp. An annual spring cleaning of Shetland's roadsides and beaches, involving over a thousand volunteers, was organised—the Voar Redd-Up, Spring tidy-up, in dialect; and the scrapyard was given council subsidies for machinery to smash the old cars into tiny pieces of metal which could be sorted and baled into saleable bundles. In one year alone Shetland exported some 4,000 tonnes of scrap metal. A new industry had been born, spawned by the oily prosperity which had caused all the trouble in the first place.

Shetlanders, who for so long had moaned about the mote in the oil industry's eye, were at last doing something about the beam in their own. The amenity trust has begun to restore Shetland to its original state, even planting native trees, which used to grow here before the Iron Age settlers wrecked the natural environment with their sheep. The trust is now a popular and well-supported body, after meeting fierce resistance at first from some councillors who regarded it as a waste of money. Attitudes have changed. A sizeable minority of Shetlanders, perhaps even a bare majority, now regard the trash as a problem that must be dealt with. This has come about through persistent nagging in the local press and local radio, often by people who, like Mr Nickerson, are not native Shetlanders but appear to care more about the place than some who are. The international concern about and publicity for "Green" issues has certainly helped. And perhaps those indignant tourists who wrote in all those years ago to berate their slovenly hosts about the "Shetland Roses", that is McEwans' distinctive red beer tins, in the road ditches have had some effect as well.

There is a lot to be done: you can still drop Shetland Roses and candy wrappers in Commercial Street, Lerwick, with impunity; the police do not have the time or the staff to enforce the litter laws; old cars are still piling up outside some crofts—in the island of Yell there is a new fashion for lining them up in neat rows; not every local fishing boat cook has got the Government-sponsored message that "Over the Side is Over" and still tips his plastic and cans over the gunwale as he passes the Bressay lighthouse; weekend do-it-yourself motor mechanics still flush sump oil down the drains, only now they do it in the dark; Lerwick still pumps its raw sewage, used sanitary towels and prophylactics straight into the sea (prompting more letters to the press from tourists who have been unwise enough to bathe in it); the poisonous plastic smoke from the council's low-temperature refuse incinerator still blankets the town in northerly winds; a few fast food shops still use CFC foam trays; gardening shops still sell mercury and dichlorvos, the latter being used extensively by the salmon farming industry to "control" sea lice on their fish; crofters, those cult figures of the wonderful Shetland traditional way of life, still leave plastic bags and sheets lying around to choke their cattle; they still poison trout streams with dichlorvos when they dump their used sheep-dip; fishermen still dump off-cuts of polypropylene netting in the sea, there to ensnare and kill seals and seabirds; and the Government-subsidised rape of the fish stocks and the overgrazing of the hills continues, only slightly abated. But I like to think that the Shetland Amenity Trust and its modest successes are a sign that Shetlanders are beginning to understand that it is not just the oil industry that can threaten the environment.

Canadian settler Rick Nickerson urges islanders: "Dunna Chuck Bruk" (Don't Dump Litter) during a council-sponsored campaign to clean up the plastic effluent of oily affluence.
Photo: Malcolm Younger

An Environmental Impact Assessment 3

Whenever the oil industry arrives somewhere new, there is usually uproar among those citizens who like wild places and wildlife and fear that both are about to be defiled. One of the first tasks of the public relations men is to find out how many of these people there are living locally, what support they have from their neighbours, and what sort of fuss is likely from their friends and sympathisers in high places.

It was like that in Alaska, where the fishermen and the "cheechako greenies" (incomer environmentalists) raised hell about the routing of the Alaska pipeline and went as far as Capitol Hill to try to make them extend it down through Canada where it could join up with the main US pipeline networks across the border. If the industry was smart enough to lay a heated pipe across 800 miles of permafrosted tundra and mountains, they reckoned, it was smart enough to go the whole way and leave their fishing grounds and playgrounds free from the threat of oil pollution.

They failed, thanks to the casting vote of Vice-President Spiro T. Agnew, and the pipe was laid only as far as Valdez, saving a whole lot of dollars. There, as in Shetland, a mighty oil terminal was built so that tankers could ship the oil right on down to Portland, San Diego and beyond. The adverse environmental effects of Spiro T. Agnew's ignorance were apparent in Alaska long before the Good Friday oilspill in Prince William Sound on 24th March 1989. There was plenty of public uproar, in that peculiarly engaging, populist and forthright Alaskan style, but it got them nothing more than a fat file of oil company assurances that turned out to be worthless.

In Shetland, which found itself hosting the oil industry at the same time as Alaska, the bland assurances were not long in coming. Eric Owen, a prosperous-looking Englishman from Shell UK's PR department in London, set himself up in a little office on Commercial Street, Lerwick, in 1974. Although apparently an Anglican in religious matters, he promptly joined the local Church of Scotland, perhaps hoping to wheedle his wily, tweed-jacketed way into the confidence of the local dignitaries. He quickly found out that his employers might have a fight on their hands to rival the opposition in Alaska.

The main difference was that most of the Alaskan "sourdoughs", old-timer settlers, and native peoples wanted oil. Oil was jobs and jobs were just what they did not have enough of. But in Shetland, Mr Owen discovered after a brief scan of the local press, they had plenty of jobs and a sizeable minority reckoned they could get along nicely without this cuckoo in the nest.

Then there were the Greenies; there were hundreds of them. Ever since the mid 19th century, when the wealthy Anglo-Scottish gentry had discovered the islands' sensational trout fishing, idyllic landscape and peaceful "way of life", Shetland had been building up a useful bank-balance of friends in high places. They might have hopelessly romantic views about the place and might only spend two weeks a year there, but they were not going to sit idly by and see their idyll raped. Since the opening in 1948 of the world-famous bird observatory in Fair Isle, mid-way between Orkney and Shetland, the gentry had been joined in their enthusiasm for Shetland by legions of birdwatchers, many of them from somewhat lower orders of society, whose idea of fun was to dress up in oilskins and tramp around muddy fields in the middle of March or November, checking off rare birds in the bulky lists which they carried at all times. These were the "twitchers", so called from their instinctive behaviour whenever news reached them of a new rarity alighting in Shetland. By the late 1980s, transporting, feeding and accommodating the twitchers had become a sizeable Shetland industry in their own right. The binoculared friends of Shetland were numerous and vocal; many of them returned to the islands in the summer to ogle the sensational breeding colonies of seabirds, some of which were also satisfyingly rare.

Worse still, from the point of view of Shell's accountants, a lot of local people had an intelligent and caring interest in wildlife, while not being above shooting the larger rarities for the pot or to prevent them mauling their smaller livestock (the sea eagle, or erne, had been exterminated in the 19th century because of its unfortunate predilection for weakly lambs). They quickly formed anti-oil alliances with the new wave of young, middle-class people who had come to Shetland to follow their professions or to "get away from it all".

The Lerwick air was soon loud with protests about the real and imagined effects of an oil terminal. It must be said that many of the worst fears have proved to be illusory so far and many a well-meaning correspondent in the "Our Readers' Views" column of *The Shetland Times* went a bit over the top, the present writer included. A local branch of Friends of the Earth was set up and prophesied wholesale oil pollution, leading to the end of Shetland civilisation as we knew it: the tankers would run aground; the oil pipelines would leak; thousands of tonnes of oil would wash ashore from undetected spillages on the production platforms; the effluent from the terminal would poison the sea and the very air we breathed (ignoring the fact that, on an average winter's day, Shetland has a complete change of air every 15 minutes, such is the force of the wind).

If it had all been confined to "Our Readers' Views", it would not have mattered. The column was not followed closely by the powers in the land. But inquisitive television reporters and hacks from the national press were starting to arrive in Shetland and publicise to a wider audience the concerns of Shetland's Greenies, ecologists, environmentalists and Friends of the Earth. Some fairly exaggerated things were being said and the aforementioned gentry were beginning to have quiet words with chaps they knew in oil company boardrooms and the corridors of the Houses of Parliament, to say nothing of the lobby of the Town Hall in Lerwick where a majority of councillors was becoming increasingly concerned about the local and national chorus of environmental concern.

The answer, as in Alaska, was to be an environmental impact assessment, produced by the Sullom Voe Environmental Advisory Group (SVEAG). This was set up in 1974 by the council and the oil industry as "a new approach to conservation in a major development"—the words used in the environmental impact assessment when it was eventually published in May 1976. Both parties, SVEAG declared, were "determined that the development at Sullom Voe should not materially harm the Shetland environment". Other interested parties were invited to join SVEAG in its secret sessions (would the Alaskans have stood for that?) and they included everyone the council and the oil industry could think of—BP, Shell, Conoco, the council, professors from the universities of Aberdeen and Dundee, the Government's Natural Environment Research Council, Marine Survey Office, Department of Trade, Scottish Development Department, Department of the Environment, Nature Conservancy Council and Field Studies Council.

Many other organisations contributed advice and comments. But the key issue which was to emerge concerned the staff who actually ran SVEAG: the chairman, Mr Peter Brackley, was a BP employee; so was the secretary, Mr Eric Cowell; the chairman and secretary of SVEAG's Working Group on Oil Spill Control also worked for BP; likewise the chairman of the Firefighting and Safety Working Groups. Only the Ornithological Working Group and the Monitoring Working Group had non-BP chairmen—respectively Professor George Dunnet of Aberdeen University and Dr Martin Holdgate of the Institute of Terrestrial Ecology (who was to become one of the Government's top environmental mandarins in the 1980s).

SVEAG was to be "essentially independent". Its stated aim was to "ensure that environmental considerations are taken into account in the planning, development and operation stages of the project". Significantly, responsibility for "environmental aspects offshore" was to remain with the oil companies. The use of language in their publicity releases was interesting: we noted the words "approach", "materially", "essentially", "considerations", "taken into account" and "aspects". SVEAG had, in the final analysis, no teeth. "The role of the group will be advisory," they said. It was to be paid for by the Sullom Voe Association (SVA), the joint oil industry/council body nominally responsible for running the show at Sullom Voe. And it would have to seek the permission of the SVA "or other appropriate body", that is, the Government or the oil companies, for expenditure on "investigations and services".

Never mind, the emergence of the environmental advisory group mollified the protestors a bit. At least their concerns were being investigated and SVEAG at first had every appearance of being an impartial body whose advice, the industry said, would be carefully considered. So far, so good. The full implications were not revealed until SVEAG published its environmental impact assessment and, in early June 1976, launched it on a sceptical audience in a "seminar", held in the cinema of the oilworkers' construction camp at Firth, a couple of miles down the road from Sullom Voe. The industry might just have got away with publishing its report. It did not get away unscathed from the seminar.

Two hundred people attended. It was an organisational disaster from the start: the bus tours of the terminal site were put on after the meeting had ended—so that many in the audience were unable to raise questions which would have occurred to them, had they visited the site beforehand. There was little time allowed for questions and discussion in any case as most of the day was taken up with lectures by contributors to the environmental impact assessment—many of whom merely summarised what they had written in the report. This was perhaps just as well, because most delegates did not receive their copies of the report until they entered the hall and only the press and the council had been favoured with advance copies.

All this created a mood of resentment in the audience—a feeling which was exacerbated when speakers repeatedly evaded questions, equivocating when it would have been more honest to say that there were some things they simply did not know. And it was noted that the publishers of the report, the Thuleprint printing co-operative in Sandwick, had printed a disclaimer on the title page, emphasising that the views it contained were those of SVEAG and its working groups, "not necessarily those of the publishers". The reason for this was that Thuleprint, although nominally a co-operative, was in fact controlled by one man, John Button, who also happened to control the local Friends of the Earth branch which he had founded and which ceased to exist when he subsequently left Shetland to join the Findhorn "alternative community" in Morayshire on the Scottish mainland.

Ian Clark, the council's chief executive, had seen the SVEAG report. He did not like it at all. So intense was his dislike that he took the unusual step of inviting the *Shetland Times* reporter assigned to the story (me) to his home in Lerwick one evening and briefing me for several hours on exactly how and why he regarded it as a biased and unacceptable report.

Mr Clark's displeasure appeared to have percolated down from the Town Hall, for not a single senior council officer attended the seminar at Firth in an official capacity. Whether this was under orders or not I cannot say, but it had the effect of a very public and pointed boycott. Ian Clark had already taken steps to dismantle SVEAG, which he clearly regarded as being far too close to BP, and to set up a rather different body in its place.

The affair at Firth was therefore not so much a learned symposium as a memorial service for the industry's first, failed attempt at environmental public relations. There was no doubt that the BP staff who worked on the report meant well. The report was a fairly comprehensive description of what Shetland's environment was like and how vulnerable it was to even minor oil pollution. They had done their homework and had come to the meeting with a list of other things they would like to research.

A central point of conflict, surprisingly enough, was a proposal to count all of Shetland's seabirds. George Dunnet, professor of zoology at Aberdeen University, (a department which was subsequently to gain a number of oil-related research contracts, it must be said), put the scientific point of view: it was not enough to say that there were lots of seabirds in Shetland—they needed to know as exactly as possible how many there were before the oil terminal opened, so that if anything happened afterwards they would know how bad the damage had been. And that meant counting them, or "monitoring", as the scientists put it, to make it sound more important. In a possibly rather unkind turn of phrase in my subsequent *Shetland Times* article, I was to define monitoring as "a word of academic origin, meaning a long holiday at the public expense". John Scott, the farmer who owned the island of Noss, a National Nature Reserve famous for its teeming thousands of breeding seabirds, put it more politely at the seminar: "If you count every bird in Shetland, what are you going to do with the information? It won't stop oil spills".

This was precisely the point. It might be very interesting to have "grown men belting up and down Yell Sound in inflatable boats, counting eider ducks", as another of my infelicitous phrases put it, and the results, displayed in some poorly-designed maps in the SVEAG report, might be of great use in improving our knowledge of the biology and population dynamics of the seabird colonies—but what had this to do with designing an oil terminal which would spill as little oil as humanly possible?

There is an apocryphal local story that, some time before the Firth seminar, a Shetland fishing boat had been steaming up Yell Sound one night when the crew felt a bump and noticed that they were no longer making any way through the water. A man was sent to the bows and returned with the surprising news that they were steaming against the Shell Brent pipeline which had been laid across the sound shortly before.

Whatever the truth of the story, the fact is that the concrete-coated pipe, at that time empty, did indeed float to the surface. Massive concrete saddles were quickly made and lowered onto the pipeline, restoring it to its designed position on the seabed. Large quantities of gravel were dumped on top of it. This did the trick; it has not surfaced again and, so far, it has not leaked any oil. But for the oil industry the incident was an embarrassing example of how their big technology could let them down, or up, in this case.

Curiously, the SVEAG report made no mention of this incident, which had obvious environmental aspects. Nor did it discuss the feasibility of handling very large tankers in the strong tides of Yell Sound. The environmental dangers of a single-engined ship losing power and drifting onto the rocks were obvious—perhaps too obvious for the report to discuss the pros and cons of allowing single-engined tankers, with single rudders and single, rather than double hulls, to navigate where no such vessels had ever navigated before.

There were to be several incidents where ships lost power or steering in this way but all were, thankfully, dealt with promptly and safely by the Sullom Voe tugs and their enterprising crews. But in 1976 there was a great deal of public concern in Shetland about the anticipated navigational problems. SVEAG's report did not address those concerns, and, in answer to questions at the seminar, Captain W. Lawrence, Shell's chief marine superintendent, said he was quite satisfied that the big ships would be able to use Yell Sound safely. He was right, as it turned out, but on this issue, as on many others, the public were asked to be satisfied with the assurances of an expert rather than being allowed to see the detailed evidence for themselves.

The SVEAG report was described as a "summary" of the research papers which had been prepared for the environmental impact assessment. In order to be summarised, the work must already have been done. But SVEAG made it clear that they had no intention of publishing these papers for at least a year. This did not lessen the suspicions of those who were quickly gaining the impression that the oil industry had no intention of being frank with the Shetland public on the details of its Sullom Voe plans.

A BP speaker, Mr P. D. Holmes, was more forthcoming on the industry's ability to clear up oil spills. He stressed that prevention was better than cure and there were only two ways to clean oil up once it had been spilled; one was to contain it within floating booms and the other was to spray it with dispersant chemicals. Booms did not work in high winds, nor in currents of more than one knot; and the chemicals, although less toxic than those used on the 1967 *Torrey Canyon* spill off Cornwall, could still do damage. In some environmentally-sensitive areas the only thing they could do was sit back and let Nature clean up the mess, which is precisely what was to happen with much of the *Esso Bernicia* fuel oil spill early in 1979. Mr Holmes did not mention what was to become the most familiar method of dealing with Sullom Voe's little and medium-sized spills as the years rolled on—labourers shovelling the stuff into buckets, once it had been driven ashore on "sacrificial" beaches.

Like the other oil company speakers, Mr Holmes was full of reassurance: fortunately, North Sea crude was lighter than Middle Eastern oil and would therefore evaporate more quickly. Most slicks would evaporate, disperse and break up in little more than a week.

As long as oil remained on the surface, rapid biodegradation would take place. Oil was, after all, a "natural" organic material.

It occurred to me that perhaps Mr Holmes had been the genius who thought of putting the quotation from Henry David Thoreau on the title page of the SVEAG report. It included the sentence: "Time will make the most discordant materials harmonise". Thoreau, who wrote this on 23rd June 1852, long before oil terminals had been thought of, would no doubt have been amused and horrified to hear his words uttered in the service of an industry which, it is fair to assume, he would have regarded with suspicion. Thoreau, indeed, might have been of the school of thought which holds that oil and water do not and should not mix. The quotation reinforced the impression of the seminar delegates that, if and when the really big spill happened, there was almost nothing that could be done about it. This impression was confirmed when we saw the plans for dealing with oil spills in the harbour itself: they envisaged holding equipment at Sullom Voe to handle just 2,000 tonnes of oil, bringing in back-up from the south if necessary. Obviously, this oil spill contingency plan was for the little ones. It was as near as SVEAG got to admitting the truth, which is that no-one has any idea how to deal with big spills, other than to stop them happening in the first place. The lesson was to be learned all over again in Alaska in March 1989.

For Peter Brackley, chairman of SVEAG and also director of BP's environmental control centre, the seminar must have been a depressing business. A report which should have crowned his career as a BP environmental scientist was being severely mauled by a bunch of fishermen, crofters, birdwatchers and newspaper reporters, rather than being hailed as the great achievement which he undoubtedly believed it to be.

At one point Mr Brackley stepped in to field a question for the luckless Mr Holmes. A delegate had asked if BP had worked out what would be the extra cost of building a completely clean oil terminal. No, they hadn't, Mr Brackley replied, for the very reason that they had never even considered building anything that was environmentally unsatisfactory. As an example of a clever riposte which became a hostage to fortune, this remark is worth passing down to posterity.

One of the aspects of the terminal which proved to be very unsatisfactory indeed, for almost a decade, was the plant which was supposed to clean the oily ballast water from the tankers. Captain Lawrence had told his audience at Firth that it was very desirable that ships should have "dedicated" ballast—meaning that they always carried enough water to keep them stable but kept it in separate tanks which were never used for cargo. When the oil was loaded, this clean, segregated ballast could safely be pumped over the side.

Most of the tankers using Sullom Voe in the early days would be of the old kind, where water was pumped into the dirty cargo tanks after the oil had been taken out, and had to be pumped out again when they wanted to load at the next port. Captain Lawrence explained how dedicated ballast ships would eventually become the norm under international regulations which would require all new ships to have the system. The reality was that almost every ship calling at Sullom Voe for a cargo in the first 10 years would have an urgent need to get rid of thousands of tonnes of oily water. Even in 1989, about half the ships arriving in the port still carried dirty ballast. And double-hulled ships are still rarely seen. Alaska had a similar experience.

The oily water ballast could not be put into the sea so it had to be put ashore, into a large tank farm which was built for the purpose. Before the water could be discharged into Yell Sound it had to be cleaned. There were holding tanks to let the oil rise to the surface, then sand filters and other equipment to clean it further. The trouble was that the system frequently clogged up or broke down. Speaking to a meeting of the Shetland Civic Society in Lerwick, on the evening after the Firth Camp seminar, the same Mr Holmes explained that they would reduce the oil content in the treated ballast water to between 10 and 15 parts per million (ppm)—the level recommended by SVEAG—with an absolute maximum of 25 ppm. The SVEAG report did not discuss what environmental effects this amount of oil in the water would have, although it did suggest a monitoring programme which, as the years went on, reassured us that no discernible environmental effects had been discovered among marine life near the effluent diffuser pipe in Yell Sound. The reason it did not discuss the effects was that no-one knew what they would be. So SVEAG had made the recommendation that the oil-water ratio should be 10–15 ppm, without any scientific evidence that this was environmentally acceptable. Significantly, this was the figure which oil industry engineers believed they could deliver, using "best available technology"—that is to say, the best equipment they could afford. It was an economic decision, not an environmental one.

The same went for SVEAG's preferred site for the effluent outfall—they decided on the point of Calback Ness, conveniently near to the terminal, whereas in the end, after much pressure from the council, the industry had to agree to pipe the effluent right out by Mio Ness where the tidal currents were stronger and would mix the discharge more thoroughly with the seawater in Yell Sound. Again, this was a more expensive solution.

Not until 10 years after the terminal opened was the ballast water plant working properly. At the time of writing, it routinely reduces the oil content to 5 ppm, half of what SVEAG thought necessary, and in recent years the terminal has very rarely exceeded its discharge consents. It has all cost a great deal more than anyone expected. The laboratory records of the monitoring of the effluent outfall have not been made public, of course, and it is sampled intermittently rather than continuously, but I do have the experts' word for it.

That, alas, was all that most of the audience had at Firth in June 1976. SVEAG's assurances on ballast water treatment were another reason to suspect that the oil industry's operational and financial priorities were being given undue weight by a body whose first concern was supposed to be the environment.

There were others: a lot of attention was directed by SVEAG to the big problem of what to do with the thousands of tonnes of wet peat which had to be shifted before construction work could begin. Once it had been decided to dump it behind a new sea wall and to grass it over when it had settled down, there really were not any further environmental considerations, although there was a successful campaign to stop the dumping of wet peat into a tidal inlet which rejoiced in the name of the Houb of Scatsta—a favourite place for twitchers in the bird migration season, then as now. But SVEAG went on and on about peat and what a grave problem it was.

They also were concerned about whether it would be better to store the oil underground or in surface tanks while it was waiting to be loaded into the tankers. Underground cavern storage was obviously best on environmental grounds, to say nothing of aesthetics. But it was "proved" not to be feasible because of the huge costs of building sealed chambers in the remarkably fractured rock of Calback Ness. And much was made in the SVEAG report about the fact that, in the event of a major disaster, cavern storage would be out of commission for much longer than more easily-repaired storage tanks above ground—hardly an environmental concern. The environmental dangers of cavern storage—gas leaks to the atmosphere and oil leaks into the surrounding geology, were mentioned only superficially. The rocks were fractured, by the way, because the oil terminal was being built on top of a geological fault, although thankfully not a very active one. So the decision was made, in a secret session of the Sullom Voe Association, to recommend surface tanks.

SVEAG had also recommended "bunds" around the oil tanks—huge earthen walls designed to contain the oil if there were a catastrophic leak. On environmental grounds one would have expected them to opt for a bund around each tank, big enough to contain the entire contents of that tank in the event of a rupture, plus a small safety margin to hold the stuff if it slopped around inside the bund. In fact, they agreed to two tanks being put into each bund but the capacity of the bund was not to be two tankfuls; it was to be one and a bit. The argument was that both tanks would not often be full at the same time and that they could not envisage the entire contents of both tanks being lost. Sullom Voe was built with 16 tanks but with bunding to dam up the contents of only about 10. This was a cheap solution to another BP budget problem, not an impartial recommendation on what was best for the environment.

Amazingly, the report went on to say very little at all about the environmental hazards of handling the huge quantities of liquefied petroleum gas (LPG) that the terminal was to produce. The location of the LPG tanks was not even shown on SVEAG's maps. Sullom Voe is firestorm country and a sizeable part of northern Shetland would be singed if the gas tanks blew up. But SVEAG did not even speculate on the ecological effects of roasting the ravens who were to take up residence on the terminal's "farm" of tanks. The birds liked the warmth and nested successfully, oblivious to the fact that the roofs of the tanks floated up and down on top of the oil.

There were other examples of the balance sheet getting tangled up with the disinterested scientific approach. They need not detain us at this late date. But the cumulative effect of this whitewash job was to make councillors extremely suspicious. There was no opposition to Ian Clark's demand that the council should withdraw from SVEAG. SVEAG's senior BP office-bearers, Peter Brackley and Eric Cowell, resigned within a week of the Firth fiasco. The environmentalists had won round one of the contest. After a decent interval, and with many expressions of mutual regard, the council and the oil industry made an announcement: the Sullom Voe Environmental Advisory Group (SVEAG) was to be superseded by a new outfit, the Shetland Oil Terminal Environmental Advisory Group (SOTEAG). This time there was a more identifiably independent chairman, Professor Dunnet of Aberdeen University, and the membership was broadened to include the Shetland Bird Club, the Shetland Fishermen's Association and other bodies. It continued to meet in secret but there were no more accusations of being in the industry's pocket, although from time to time some of us have suspected it of being less than willing to rock the boat when a good rocking was exactly what the boat required. But, in general, SOTEAG, son of SVEAG, has done a much better job than its industry-controlled predecessor, and public confidence has very largely been restored. Its acronym is easier to say, too.

And all that monitoring has been done. Thanks to Sullom Voe, we now have a much better idea of how many birds and animals share these islands with the inhabitants and how they make a living from the natural environment which humans, although not particularly the oil industry, have altered so much. The numbers of oiled seabirds found during the monthly "beached bird surveys" of selected Shetland beaches have declined dramatically since the early days of Sullom Voe (to which we shall return in Chapter Five).

Ironically, much of the data on seabirds has been put to good use in campaigns to force the fishing industry to adopt more environmentally sound practices and stop killing the small fish which feed the seabirds. That is not to say that the fishermen's concern about the oil industry was unfounded. There was good reason to be worried in 1972–76 and beyond. But, through a combination of good luck, media pressure and, above all, the highest professional standards at the port of Sullom Voe, the worst has not yet happened. To borrow a phrase from Churchill, the price of relative freedom from pollution has been the need for eternal vigilance.

The lesson for anyone courted by the oil industry is—beware the environmental impact assessment. It may seem like a good idea, but there are a number of disadvantages evident from the Shetland experience of SVEAG and SOTEAG:

- The assessment and its surveys will defuse public opposition to the oil development, usually at a crucial stage of the planning process;
- Unless the local politicians are up to snuff, surveys like this will be used to delay key decisions on environmental safeguards until the developer has had a chance to work on the waverers. This did not happen in Shetland because of one determined wise-guy called Ian Clark, who helped the local politicians to keep their nerve and face down the industry on the SVEAG affair, but it could easily have happened without him;
- Commonsense observations by local people—such as the universal and sensible opinion that oil is bad for birds and it is best to keep them apart—will be derided by the so-called professionals who will demand numbers and then waste

months counting critters when everybody should be mounting campaigns;

- Environmental scientists, always short of cash for their pet projects, will jump on the bandwagon and use the survey to payroll all sorts of basic science which is unnecessary for the purposes of preventing oil pollution—we do not need to devise a logarithmic scale showing exactly how bad oil pollution is for seabirds. It is always disastrous;
- If there are a lot of birds you can see them. We do not need environmental impact assessments, telling us exactly how many, for us to notice when they start to decline. And if the dead ones have oil on them you do not need to be a research scientist to work out why;
- Unless the body carrying out the survey is rigorously independent of the parties commissioning it, and chaired by someone who has no financial interest, other than his or her fee, you can end up with the conflicts of interest which surfaced in SVEAG's report—the technical solutions which are best for the developer's bankbook are confused with and can override those which are best for the environment;
- In this context, expressions such as "best available affordable technology" usually mean: "OK, you guys, this is a load of second-rate, cut-price gizmos which we're gonna offload on you under cover of some public relations because we figure you don't have anyone smart enough to figure out what we're doing". Go on—figure it out; spoil their week;
- Local energies should be concentrated on dragging the developer kicking and screaming to the cleanest and greenest solution in sight, if necessary by using the courts. He will threaten to abandon the development if things get too green, that is, expensive. These threats are usually bluff and should be treated as such. If there is oil out there, he wants his hands on it and almost anything you can get out of him will be negligible to him, however much it seems to you;
- Just as important as all that monitoring is an effective system of policing what the developer has agreed to do. If they fail to come up to standard, make them do it again. No compromises on environmental safeguards and, by the way, make sure you know the difference between "continuous" and "continual" when it comes to sampling and testing effluent discharges—

these people have word engineers as well as pipeline engineers;

- No private deals and no secret committees—go public, Joe Public, or the developer will bind your representatives hand and foot with "commercial confidentiality".

And that's all I have to say on the subject of environmental impact assessments—until the next time.

Clean enough for otters—this novel roadsign warns drivers approaching the Sullom Voe terminal not to run over the otters who fish around the tanker jetties. Ronnie Gallagher, BP's energetic environmental officer at the terminal, poses for a publicity picture.
Photo: Malcolm Younger

Building the Beast 4

Two thousand years ago, an unknown Celtic genius in Northumbria invented a piece of technology which revolutionised the energy economics of his tribe.

It was a special spade, designed for cutting peat into handy-sized lumps which could be laid out to dry in the sun and the wind. After a month or two, if the weather was fine, the stuff was dry enough to burn. No need to chop down trees, except for kindling, and no need to grovel around in damp holes hacking out lumps of coal. It had half the calorific value of coal, too, although our unknown Celtic genius would not have put it like that.

How do I know all this? Well, the earliest known version of the peat spade was found in a Northumbrian bog and dated to about the time of Christ. There may have been earlier ones, but if there were they are still buried in a bog some place. Shetland, like Northumbria, is well provided with peat bogs and moors, but, even so, it seems to have taken about 800 years for the peat spade to make it this far north. This may not have been entirely due to native conservatism and a distrust of the new-fangled; there is evidence, in the form of tree roots and branches found in the valley bogs of Shetland, that these now almost treeless islands once supported a widespread low scrub of willow, birch, pine, rowan and hazel. This would have kept the early settlers in firewood for many hundreds of years until it was all cut down and their heather burning and sheep prevented natural regrowth of this archaic woodland. If you want to see the last remaining hazel "tree" in Shetland, it is four feet high and can be found in a little ravine at Catfirth, just off the road from Lerwick to Sullom Voe.

The resort to peat was perhaps a response to a shortage of more easily-obtained fuel. A Norse earl by the name of Torf Einar, Einar the Turf, is traditionally credited with introducing the Viking version of the implement, the *torf-skaer*, to the far north. Here, we like to think, it was refined and developed to perfection as the inimitable Shetland *tushkar*. Local blacksmiths make them to this day, and there are still learned arguments going on about the best design.

In Ireland, the Western Isles and Orkney, those other parts of the old Viking empire, they have tushkars too, or something bashed together to look like them. But, we firmly believe, their technology is crude; the slab of peat is usually cut from the wet moor and dumped down on the wet grass at the lower side of the peat bank's cutting face where it will never dry. Then some poor creature, almost always a woman, has to wrestle with the great slobby slab of mud and heave it up onto the higher side of the bank, where it may dry out in God's good time.

In Shetland, on the other hand, the *tushkar* is so designed that the slab is cut, lifted and neatly laid on the top, dry, side—all in one graceful movement. That is the theory at least, although it takes a while for some of the newcomers to master the art. Women still raise the peats into little pyramids once they have dried a bit and turn the damp sides to the sun when the time comes for the final drying process. Cutting and drying the peat is a back-breaking business which can take from April to August in a bad summer. There is a Shetland joke that all the heat from peats is from the working of them. As a peat cutter myself, I can testify that this is not really a very funny joke. The writing of this book, I am glad to say, prevented me cutting any peats in the summer of 1989. I was not sorry at the time, but as winter drew on I was having second thoughts.

In some districts of Shetland they have gone in for mechanised peat cutting, sometimes on a commercial basis. Dried peats have even been exported to Faroe. But in my home island of Bressay, where the islanders once cut peat to sell to Lerwick (an illegal trade, frowned upon by the 18th century landowners), this modern innovation has been resisted on the grounds that it ruins the pasture. Proper peat cutting improves the pasture because the green sward is laid down again on the clay from which the year's peat harvest has been taken and, over the centuries, large areas have been improved in this way.

At Calback Ness in Delting, overlooking the once tranquil inlet of Sullom Voe, the crofters cut their peats for a dozen centuries or so before they gave it up for a bad job and moved to more salubrious parts of Shetland. Whatever else persuaded them to flit, it certainly was not a peat famine, unlike other areas of Shetland, such as the island of Papa Stour and the Easting district of Unst, where the peat had been exhausted by the 18th century. There would have been enough peat in Calback to keep the fires going for another couple of thousand years at least.

These huge reserves of one fossil fuel were to be the oil industry's first big problem when their pioneers arrived at Calback Ness in the summer of 1974, to work out how to make the place fit to receive a fossil fuel with a rather higher calorific value. There was no road into Calback Ness then—the surveyors had to drive all-terrain vehicles along the bouldery beach from the asphalt at the Sullom Voe Hotel, a former RAF officers' mess at Graven, a spot as bleak as its name, one mile to the south.

The surveyors stuck rods down into the peat but could not find bottom. They brought drilling rigs along the beach, by now levelled with bulldozers, and took core samples from all over the site. They discovered that the peat was 15 feet deep in places. And there was underlying rock to be dug out too. They tried to estimate how much of peat there was. It could only be a guesstimate. By the time the last load of peat was shifted, over three years later, BP calculated that some 10 million cubic metres of rock and peat had been moved—and that was only half of the surface covering of Calback Ness.

If anyone ever has to eke out a living there again, when the oil terminal has faded into folklore, there will be plenty left to burn as the bards of the 23rd century tell tales around the peat fire of the lost age of black gold.

But I am waxing poetical and we have a prosaic but nonetheless heroic tale to tell. For the pioneer crews at Sullom Voe in 1974, life could be extremely rough. The luxurious construction camps were still on the drawing board. Just to peg out the site was one of the biggest surveying jobs ever done. The deep, heathery hill of Calback Ness, pitted with bogs and pools, was hard going for men on foot with levels and poles. The headland looks small on the map but it is a big place when you have to walk it. There was no shelter from the violent rain squalls which can lash Shetland without warning, even in the "summer". But for two months a year it was never properly dark, so the work could and did go on at all hours. Oilskins and rubber boots were everyday wear. Dormitories were set up in the Sullom Voe Hotel, formerly the quiet retreat of fanatical trout fishermen on their Shetland holidays. The low, concrete building became the site's first social centre. There were only a few dozen people there in the early days and everyone knew everyone else in the bleak bar, made cheerful by the genial host, a former Edinburgh

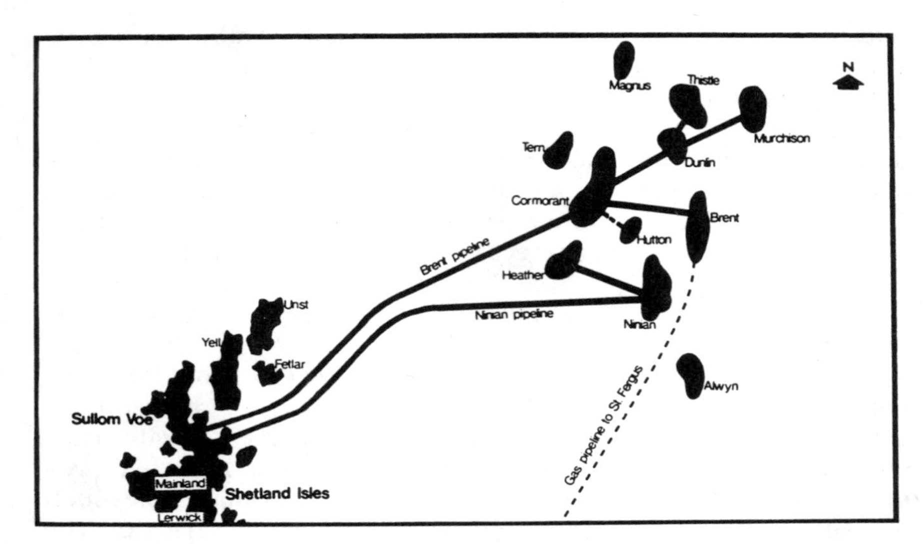

Sea routes of the two 36-inch diameter pipelines from the Ninian and Brent Group oilfield to the Shetland Islands.

Credit: Reproduced from *Shetland's Oil Era* by Elizabeth Marshall with permission of the publishers, The Research and Development Department of Shetland Islands Council. Original map courtesy of BP.

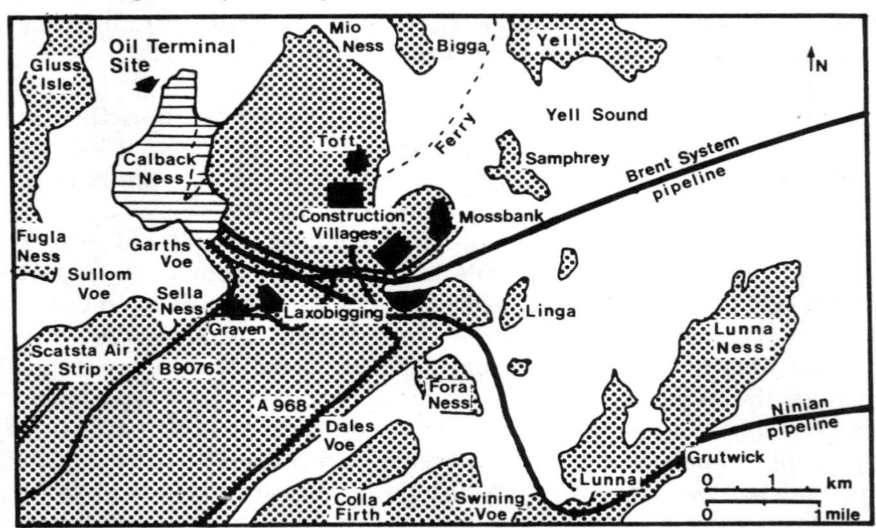

Oil Terminal Site at Sullom Voe and Pipeline Routes.

Cartography: Donald Battcock, from a map published in *Shetland's Oil Era* by Elizabeth Marshall, published by The Research Development Department of Shetland Islands Council. Original map courtesy of BP.

schoolteacher called Dave MacDonald, whose enthusiasms included a bizarre collection of neckties on display among the whisky bottles.

Bedspaces were scarce. Even the semi-derelict, 19th century estate factor's house at Garth, overlooking the beach road into Calback Ness, was pressed into temporary service as a bunkhouse for some of the pioneers. But, despite the privations, there was a real community spirit—which was to dissipate somewhat as the workforce built up to its amazing peak of 7,180 people in November 1980.

Even that figure does not convey the magnitude of the project—in all, some 25,000 people were involved in designing, building and commissioning the terminal. And then there were the people on the mainland of Britain who made all the bits of heavy ironmongery which had to be shipped north and bolted down—they probably brought the total numbers involved to something nearer 50,000. Sullom Voe was a major stimulus to the British economy even before the oil and its revenues began to flow; and it meant almost as much to the job-starved, heavy industrial rivers of the Clyde, the Tyne, the Forth and the Mersey as it did to Shetland.

The basic surveys had already been done from existing maps and aerial photography and, even as the foot soldiers got to work, a more comfortably-housed army of planners and designers was at work in London and Aberdeen, making the detailed plans which had to be pegged out on site. One of the organisational problems of building Sullom Voe was making sure that while the survey work went on, the designs kept coming in the right order, the earth-movers were ready at the right time, and the heavy plant was being prefabricated all over Britain, and elsewhere, so that it would slot into the jigsaw when the time came. The logistics of it can be compared with the Normandy landings and, life being what it is, there were occasions when huge lumps of machinery arrived long before (and sometimes, worse still, long after) their foundations had been built. But there were many logistic triumphs among the foul-ups, and, to this day, there is a camaraderie among the men and women who worked on the early years of the project, just as there is among the veterans of the Alaska pipeline.

Before they could make a serious attack on all that peat and rock, two things were essential—an access road and a pioneer camp. The camp went up at Firth, two miles away, in 1974; and, until there was a road from there to the site, a local man, Arthur Manson, used to ferry the workers in his boat across Garth's Voe, an inlet of the larger Sullom Voe. The men worked for a well-known contractor from the north of Ireland, JMJ Ltd, which later merged with the LJK consortium—Lilley, JMJ & Keir—who worked at Sullom Voe for seven years. Soon the accents of Ulster were mingling with the Shetland dialect in the local pubs, not that the poor fellows were allowed much time for going to pubs. These were the travelling men, heirs to the armies of navvies who had built Britain's canals, railways, airfields and motorways. They came to Shetland from Orkney, where they had just churned up the island of Flotta to prepare the site for Armand Hammer's Occidental Oil Company terminal in Scapa Flow. Flotta was a big job, but nothing like the scale of Sullom Voe.

The smaller bulldozers and tractor shovels could be brought to Lerwick by sea, unloaded and trucked north to Sullom Voe, where they left the asphalt and crawled along the recently-levelled beach to get to work. But Sullom Voe needed huge earth-movers, dozens of them, and the only way to get them there was to bring them in by landing craft. For as yet there was no Sullom Voe jetty to take the coasters and heavy-lift ships which would haul up all that prefabricated plant to be bolted into place.

Building this "construction jetty" was another priority. While some JMJ men tore away at the peat, others prepared the approaches to the jetty. Come Christmas 1974, they flew south for a well-earned break. When they came back in the New Year they found that the sea had washed away their work in a storm. It was the first of many setbacks caused by the Shetland weather but they pressed on, and by March 1975 the site had its first building—a cabin on the Hill of Calback, complete with a telephone and Sullom Voe's first oilwoman—Marabelle Murray, the JMJ secretary.

Clearing the site and laying out the oil terminal's roads and foundations took three years. A Glasgow company, Motherwell Bridge, world-famous for its muscular heavy engineering, moved up in 1975 to start work on the crude oil and ballast water tanks, surrounded by all this mud, sweat and occasional tears. Not until the summer of 1977 did the first of the mechanical contractors, Wimpeys, arrive to build on the foundations of the main processing plant. Yet within 18 months the terminal was up and running, although still far from complete. Even in May 1981, when the Queen arrived to perform the "inauguration ceremony", albeit 30 months after the first oil flowed, there were still 500 construction workers there, moving yet more peat, laying concrete, tarring roads and landscaping the site. Seven years after work began, Sullom Voe remained one of the biggest civil engineering jobs in Britain.

Her Majesty the Queen meets some of the "bears"—construction workers who had a day off in May 1981 for the royal inauguration of the Sullom Voe terminal. Photo: Malcolm Younger

To get it pumping oil in November 1978, just three and a half years after those JMJ men arrived, was an extraordinary achievement. The secret was prefabrication. The original plan had been to build 30 percent of the plant on the British mainland and 70 percent in Shetland. Once the oil companies had realised the logistical problems of building large steel structures in the vile Shetland climate, so far from the main manufacturing centres of the UK, prefabrication was the only game in town. The proportions quickly became 50-50 and, by 1978, 75 percent of the work was being done in the south. The strategy had a useful political angle when dealing with the increasingly prickly local business community—the "prefab policy" took a little of the superheated steam out of the Shetland economy and eased the traditional employers' labour famine by a marginal but useful amount.

If Sullom Voe had been built in the normal way it would have taken 14 years, not seven. Prefabrication was a logistics gamble.

No-one knew if all that stuff could be shipped north safely—but it was. They took the gamble because the economics looked attractive.

Norman Dunlop, project manager for Foster Wheeler's £500 million contracts at Sullom Voe, once did some sums: the extra costs of working in an "offshore" location like Shetland meant that for every £1 spent on a mainland worker in, say, a pipe fabrication yard, they would have had to spend £6 doing the same job on site at Sullom Voe, because they had to fly him in and out, give him a bed and feed and water him, to say nothing of the subsidised beer in the camp bars. So everything that could be made down south was—huge pieces of pipe in complicated shapes, vast pressure vessels, whole modules containing compressors, turbines and high petro-tech of all descriptions. Some of the loads weighed 460 tonnes, and the average was 250 tonnes—all shipped to the construction jetty, which for several years was one of the busiest piers in Britain, presided over by Wimpey Marine's manager, Barney Ryan. Lerwick's dockers were quickly in on the act, bussing up from town to earn money like never before. By 1978 they were working two shifts to clear the backlog in 16-hour days and shifting 30,000 tonnes of material a month across that jetty.

The Sullom Voe Hotel bar soon became ... well, crowded is putting it mildly. Even as the Firth construction camp was opened, a second one was being planned at Toft, an almost deserted crofting township by the shores of Yell Sound.

Toft had hitherto been notable only as the Mainland terminal of the Yell ferry. It also had a telephone booth—in the porch of an empty croft-house where I had once camped on a rainy night. I discovered that my great-grandmother had lived briefly at this croft as a little girl. She and her mother left the isle of Samphrey, out in Yell Sound, after a 19th century disaster which drowned most of the Samphrey menfolk within sight of their homes. Digging deeper into family history, I found that some of my ancestors had once hacked a living out of a croft on Calback Ness itself. Sullom Voe was becoming a family concern.

They built the Toft camp and still there were not enough beds—those 1972 predictions of "400 construction workers" had either been a massive miscalculation by the oil industry's forward planners or a clever piece of misinformation to beguile the local population into under-estimating the scale of what was to be. After all, they had already started work in Alaska and must have had some idea of how many people it took to do a job like Sullom Voe at 60°N.

So a big ship came into Garth's Voe, next to the site, and was securely moored with a pontoon gangplank to the shore. She was the *Rangatira*, a former New Zealand ferry converted into a floating hotel. Even then they were short of bunks. Another accommodation ship, the Swedish ferry *Stena Baltica*, was anchored nearby. And that seemed to satisfy them, although demand for beds in Shetland hotels many miles away was such that some of them had to turn away bookings from tourists—not that they could have catered for them anyway, for most of the available labour had signed on at the camps as domestics, for rather more cash than the hoteliers were willing or able to part with.

The island of Yell became a dormitory, with over 10 percent of its 1,000 population commuting across the sound to the camps and the construction site at one point. The council drew up a policy to control the trailers which Sullom Voe workers were setting down all over the north Mainland; the workers ignored edicts to move and, when the council planners became too annoying, they simply went on to another impromptu trailer park nearby—and the planning control process had to begin all over again.

One man's experience summed it all up. Jim Hallam, the safety officer for Foster Wheeler, one of the biggest Sullom Voe contractors, came north in 1974. In less that three years his address changed from the Firth pioneer camp to the Lerwick Hotel, the St Magnus Bay Hotel at Hillswick, Firth construction camp, Busta House Hotel, the Westings Hotel, a rented room in Brae and, at last, a council house in Brae's new "toytown" development.

Safety officers like Jim were kept busy. BP made a big thing of safety, often leading to ribald comments in the site's *samizdat* or underground newspapers, the *Ungam Gazette* and the *Staney Hill Free Press*, that were alternately tolerated and frowned upon by the more serious-minded BP managers. In theory, and increasingly in practice, if you did not wear your safety helmet on site you could lose your job. The BP medics in the Firth surgery did not sit around waiting for customers, but the most hardened cynic must admit that the company's safety record was above average. The rest of the British construction industry was nowhere near Sullom Voe's safety record in the first seven years, and, despite some appalling accidents, leading to deaths on several occasions, it was one of the safest construction camps in the world. That is what the safety statistics say.

Foster Wheeler and BP themselves won national safety awards, after drilling into the construction workers that they were serious about safety. The achievement was all the more remarkable because they worked punishing 12-hour shifts, seven days a week and rarely stopped for the weather. Men might be "rained off" on other British construction sites; not at Sullom Voe. This was a key part of the deal struck by the company with the big construction union, the Transport and General Workers Union, and the main electricians' and engineers' unions on site. Big, big money and those longed-for free flights for home leave every month were also central to the deal.

In the planning offices which sprawled around the mud and dust of the site, the managers were constantly beset by what came to be known as the "milestone complex". Oil company partners, contractors, the media and, above all, the cash-hungry Government, had been told of target dates for completing various bits of the huge jigsaw puzzle on Calback Ness. Office calendars were festooned with milestone dates. And people kept asking if this or that milestone had been reached yet. Often it had not. The size and complexity of the puzzle meant that any target had to be a guesstimate, at best. The weather, construction strikes—although the big unions saw to it that there were few of those—the distance from the main fabrication centres, a desperate shortage of concrete, and, in the early days, the fights between Shell and BP, and between both of them and the council, all held up the project.

The council was perhaps the least cause of delay. Once the Sullom Voe Association deal had been done in 1974, council planners bent over backwards to rush the detailed plans through the normally cumbersome British planning process. And individual councillors rarely had enough technical information to mount effective delaying tactics against plans which made them unhappy.

The Shell/BP fights took longer to settle. In the winter of 1975 the site looked, and was, a disorganised mess. That's BP's version, anyway. Different contractors were building different bits and pieces here and there, without any obvious overall control or strategy, according to Ted Ferguson, the bluff BP construction manager who arrived at Sullom Voe in 1977, fresh from his labours on the Alaska pipeline. In that year the project was reorganised by him and two other members of BP's "Alaska Mafia", as the unofficial site newspapers called them—Basil Butler and David Henderson, the senior managers who effectively ended Shell's control at Sullom Voe and made the contractors do it BP's way.

The problem facing Shell, and now BP, was that the project could not proceed in orderly stages—all the stages overlapped, so that by the time the grass was growing on some of the tank farm bunds, others were still being carved out of the peat. Building, commissioning and operating the terminal overlapped for several years, creating a running nightmare for the men who were supposed to be in charge and the harassed clerical and secretarial staff, women, mostly, who tried to keep the managers' offices together. And all the time, those ships and barges loaded with prefabricated modules for the site were steaming into Sullom Voe, jostling for a berth at the overloaded construction jetty, which also had to handle thousands of tonnes of sand and gravel for the stunning quantities of concrete which the site demanded every day.

Sand and gravel became big business in Shetland. Several local quarry owners became rich beyond their neighbours' wildest nightmares, although one or two went bust in spectacular fashion. Big quarries were opened up and invariably caused vehement controversy. The trucks rolled day and night to feed the insatiable concrete batching plants at Sullom Voe. In one nine-month period, Amalgamated Quarry Services churned out 47,000 cubic metres of concrete.

Some of the Shetland roads, which the early planners had envisaged as being ready before this extra traffic arrived, were still not completed 10 years after the terminal opened. Much of the north and central Mainland of Shetland became scarred, with various beauty spots despoiled by insensitive quarrying. Private enterprise did its bit, but the council caused some environmental outrages with its own quarries for roadstone and concrete—notably at Mavis Grind, the celebrated isthmus where it is possible to throw a stone from the North Sea to the Atlantic.

Most of the construction workers came from, or through, the Glasgow area. These were the "bears", as they proudly described themselves, recruited through a special training and induction centre set up at Bellshill just outside that city. They were flown 400 miles north to Sumburgh airport, at the southern tip of Shetland. The airport had to cope simultaneously with all the demands of helicopter traffic to the offshore rigs and the production platforms then under construction. For years it was chaotic. Convoys of buses bumped the weary bears, often hung over from celebrations in Glasgow the night before, over the 50 miles of pitted, narrow roads between Sumburgh and the work camps. It was never a pleasant journey and in a winter gale with horizontal sleet it could be tiresome in the extreme. Not until 1978, four years after work began, was Scatsta airport opened, right next to Sullom Voe, and they could be flown in direct. David Henderson, the senior project manager at Sullom Voe, ruefully admitted in 1981 that if he could have started all over again he would have built the new Scatsta runway first, avoiding the Sumburgh bottleneck and all the frustration, delay and extra expense it caused.

Once on the site, it was work, work, work. The pace was frantic—too frantic, as time would tell. But the unions on the all-union site accepted it and, after a strike or two, managed to get their members longer leave and shorter spells in Shetland. But it was still 12-hour shifts, seven days a week during their four-week tours of duty.

Ted Ferguson, the BP construction manager, admitted that he was not getting the stated minimum of 65 hours hard labour out of the men on site. It was more like 40 hours a week when the weather was bad, as it often was. So why the 12-hour shift when, on his own admission, people were often not working but waiting to be told where to work, what to do, or were just dodging off? The answer he gave me in 1981 was revealing about BP's rather paternalistic attitude to the "bears": "Do you think men really want to come up here and work 40 hours a week? There's got to be an incentive to work away from your family in a harsh environment in remote areas. You've given up something, even when you're sitting in the camp there with your beer, your telly and your bingo. You may come from Grangemouth or Torquay or wherever, but you're missing home and, above all, your family. So why not work?

"I think if you walk around this site you'll be surprised at the spirit", he continued. "Never mind the job's coming to an end, people are not dismayed by being at Sullom Voe. By and large, it's been a very happy construction site. If you speak to people who tour construction sites they will tell you that if you walk onto a site down on the mainland which is a 40 hour a week job, barely anyone will smile at you, let alone say good morning. I walked through the process area the other day, and there was an LJK labourer who asked me to 'spell' [relieve] him for a couple of minutes. That's the sort of atmosphere. A man can exchange a word with you. Not every site in Britain's like that."

Having a "captive" workforce helped managers like Ted Ferguson, who honestly believed that the divisions between managers and workers were to some extent being broken down. There was friction with the unions from time to time, but less than 1 percent of the available man hours were lost through industrial action. He explained: "As time has gone on I've found it easier to deal with the unions on site. It helps if you get to know the people concerned on a social level, and we now have communications in all directions. The problems occur when someone goes outside the established scheme of communications and you miss a trick somewhere. It's got easier because all the parties know each other and they know what to expect."

The biggest paint job in Europe—one of hundreds of men who painted the miles of pipework at Sullom Voe, at work in the summer of 1980.
Photo: Jonathan Wills

A giant pipe-laying barge passes Sumburgh Head lighthouse after laying oil lines linking new fields to the Sullom Voe terminal.
Photo: Malcolm Bradbury

What they could expect was laid down in the Sullom Voe site agreement—drawn up by BP after a big strike in May 1977, when it was found that the unions and individual contractors were doing individual deals—with the result that some men were getting better pay and conditions than others, who felt resentful.

"When I came here we expected a multitude of contractors to sit together and form a common policy to present a united front to trade unions", Ferguson said. "It's never happened before, so why should we think it was going to happen here?" What did happen was that BP knocked the contractors' heads together and took over the negotiations with the unions itself. It worked, as far as official strikes were concerned, but the site still had occasional, unofficial disputes, led by local union officials, "shop stewards", elected by the workers.

"The minute the shop stewards want to start negotiating themselves, without the full-time [unelected] officials, then we cut them off", Ted Ferguson said. "We can't give recognition in terms of a negotiating committee, but we certainly give them the respect that a site committee demands."

Respect—it sounded good; but in practice this meant that once the site agreement had been signed with those unelected full-time officials, bureaucrats based in union offices on the British mainland, then the unions became the company's enforcers as well as the workers' representatives. Ted Ferguson elaborated on his favourite theme: "We don't make deals with the men. We make deals with the unions. The men are the employees of a company. It is the manager's job to ensure that the men adhere to the agreement. It is the job of the shop stewards to make sure that the companies keep to the terms and conditions laid down in the agreement. They are the local policemen, if you like. If a shop steward believes that a company has breached an agreement, then I will investigate. I have done. Often. But the shop stewards cannot negotiate fundamental changes in the agreement. That can only be done by the full-time union officials."

To the shop stewards, and to many of the men, it sometimes seemed that their union officials were policing them as well as the employers. Being a shop steward at Sullom Voe could be a frustrating business. The common front of the employers and their refusal to negotiate with the men on the job meant that resentment was directed away from the individual contractors and into internecine strife within the unions themselves. It was a neat fix which cleverly exploited the internal contradictions of some of Britain's least democratic unions. Maybe they deserved all they got. As far as the locals were concerned, it took years for BP and the unions to negotiate a deal which gave a flexible shift system for Shetland residents, so that they could look after their crofts and lead some sort of family life.

All I can say is that the union bosses who visited Sullom Voe during the construction years generally seemed to be on terms of unhealthy conviviality with the BP managers and I am glad they were not representing my interests. Ted Ferguson knew exactly what he was asking the full-time union officials to do. In a tribute to them in May 1981, he told me: "It takes a lot of courage to do what the full-time union officials do at times and I couldn't do their job". One of them was a man called Tommy Lafferty, who had put "tremendous personal efforts" into Sullom Voe: "You can see the marks on him, quite honestly".

There were tensions within the management as well; BP's traditional role on a big project had been to hire the contractors and let them get on with it, not only in dealing with the unions but also in finishing their part of the job in their own way and, it sometimes appeared, in their own time. The revolution that Ted Ferguson brought to Sullom Voe was to insist on an overall plan and to co-ordinate it with an attention to detail that some companies found unnerving. His Sunday morning "prayer meetings", where all the contractors' site managers were required to meet in his office to discuss the week's progress, or lack of it, made him almost as unpopular with them as with some of the shop stewards. But they got the job done and Ted moved on to greener pastures, credited with saving BP's bacon at Sullom Voe.

The BP managers were smart, but not that smart. In 1981, when the terminal was "inaugurated" with much royal pageantry and a media circus, they had made a big thing about their quality control. A whole section of the site offices was devoted to making sure not only that the job got done but that it got done properly. But in one crucial section they fouled up badly.

In that *Shetland Times* interview with me in May 1981, shortly before he left Shetland, Ted Ferguson spoke about the "tremendous spirit of co-operation" on the site. The result, he boasted, was that the quality of the job was "first class": "I've seen work all over the world and you won't find a lot better than this", he said.

First class it was not. Six years later, Ted Ferguson was far away from Sullom Voe when the failings of his quality controllers came home to roost. It turned out that BP inspectors had accepted miles of pipework which, within six Shetland winters, had become so badly corroded that it had to be stripped out and replaced. The scale of the problem became apparent in the summer of 1987, when there were several potentially disastrous gas leaks from pipes which were found to have rusted through. The two year repair job, starting in October 1987, cost BP and partners some £75 million (they admit to £70 million)—plus the lost value of gas which had to be flared off at the North Sea production platforms, rather than being pumped to Sullom Voe with the oil. That was profit going up in smoke.

The worst corrosion was found in pipes which were lagged to keep the heat in (or out). On top of this insulation material was metal cladding which was supposed to keep the wet out. In fact, much of it was badly-made and wrongly installed, with the result that it let and kept moisture in. If the lagging had not been wet when it was fitted, it soon became so. The salt-laden Shetland air did the rest, assisted by sulphate-reducing bacteria (SRB) from the oil, which became more of a problem as the years rolled on and actually ate out the steelwork from inside. SRB is still the problem that BP does not like to talk about, onshore or offshore.

"Corrosion Under Insulation" (CUI) was the BP euphemism for the rust which spread like a cancer through the gas processing plant and much else besides. The explanation for CUI was apparently slack work by BP's own inspectors, coupled with an inexcusable ignorance of the Shetland climate on the part of the people who designed the plant in the first place. That was the down side of the Sullom Voe success story, but by 1987 the plant was making so much money that BP could easily afford the cash to make good defects which should never have happened. And no doubt the "rush job" mentality of the construction years, when any problem could be solved by throwing more men and more money at it, contributed to this hugely embarrassing foul-up. But, BP publicity men being what they are, even this was portrayed as yet another heroic technical feat by the world's most wonderful oil company and the "bears" who returned in 1987 to work those long, long hours, sometimes on the same pipes which they had last seen in 1981. First-class quality, once again, and no doubt it was, the second time around.

There had been other foul-ups. One of the key pieces of terminal equipment was a giant flare stack, 300 feet high. Without it the terminal could not accept any "live" crude from offshore—oil with the gas still mixed in it. For safety's sake, there had to be somewhere where gas could be burned off in case of a surge of pressure in the system. That somewhere was the flare stack. Without it, an escape of gas would form a cloud at ground level over the terminal and explode as soon as it met a spark.

The tower for the main stack was made in the south and shipped to Shetland in sections. While they were unloading one of the pieces they dropped it. Little damage was done, but checks showed up defects which had not been noticed before. The engineers did their sums and found that although it might stand up, it might not stand being lifted up into place. A total re-design was called for, as well as fixing some faulty welding in the steelwork of the structure. It took a year. The original manufacturer very prudently went out of business and the remedial work was done by a Lerwick engineering firm, Norscot Ltd.

During a lull in a hurricane, on 8th February 1981, the flare stack was finally levered into position—a superb piece of precision lifting by the cranes and crews of Rigging International, the company that had been waiting for over a year to do the job and who performed many other miracles on the site, along with Sunters, the company that did the heavy haulage through the mud. By nightfall it was safely guyed and bolted down. This was just as well, because a 50 knot snowstorm was blowing again by then. Only when this job was finished could the terminal start accepting gassy oil and the wasteful flaring of gas out at the oilfields was brought under control. But, as it turned out, there had also been appalling problems in persuading the gas processing plant at the terminal to work; and some of the prefabricated units arrived on site more than a year behind schedule, so it was many more months before gas processing and exports of liquefied petroleum gas could begin.

And the jetty to take the gas tankers was still not finished. Such was the rush to start shipping dead crude that one of the four tanker-loading jetties was brought into service before it was completed. Jetty #1 was used to load oil while work continued on Jetties #3 and #4. When Jetty #3 was ready, Jetty #1 was taken out of service and fitted up as the terminal's only gas jetty. Confusingly, Jetty #2 was the first to take a tanker alongside when, in October 1978, Shell's *Donovania* made a flawless trial run down Yell Sound, as had been demanded by all those worried environmentalists, back in June 1976.

The jetties were essential to BP's terminal, but building them was not BP's problem. For everything below the high water mark was built and was to be owned by Shetland Islands Council, using the oil companies' money. The idea, never fully explained, was that this would give the council some extra control over the industry, through the potential threat of closing the jetties, but, as they had signed a deal guaranteeing the oil companies exclusive use of them for as long as they liked, this seemed a tenuous sanction at best. And however bad the council/BP rows became, the threat was never carried out.

Layout of the Sullom Voe Oil Terminal

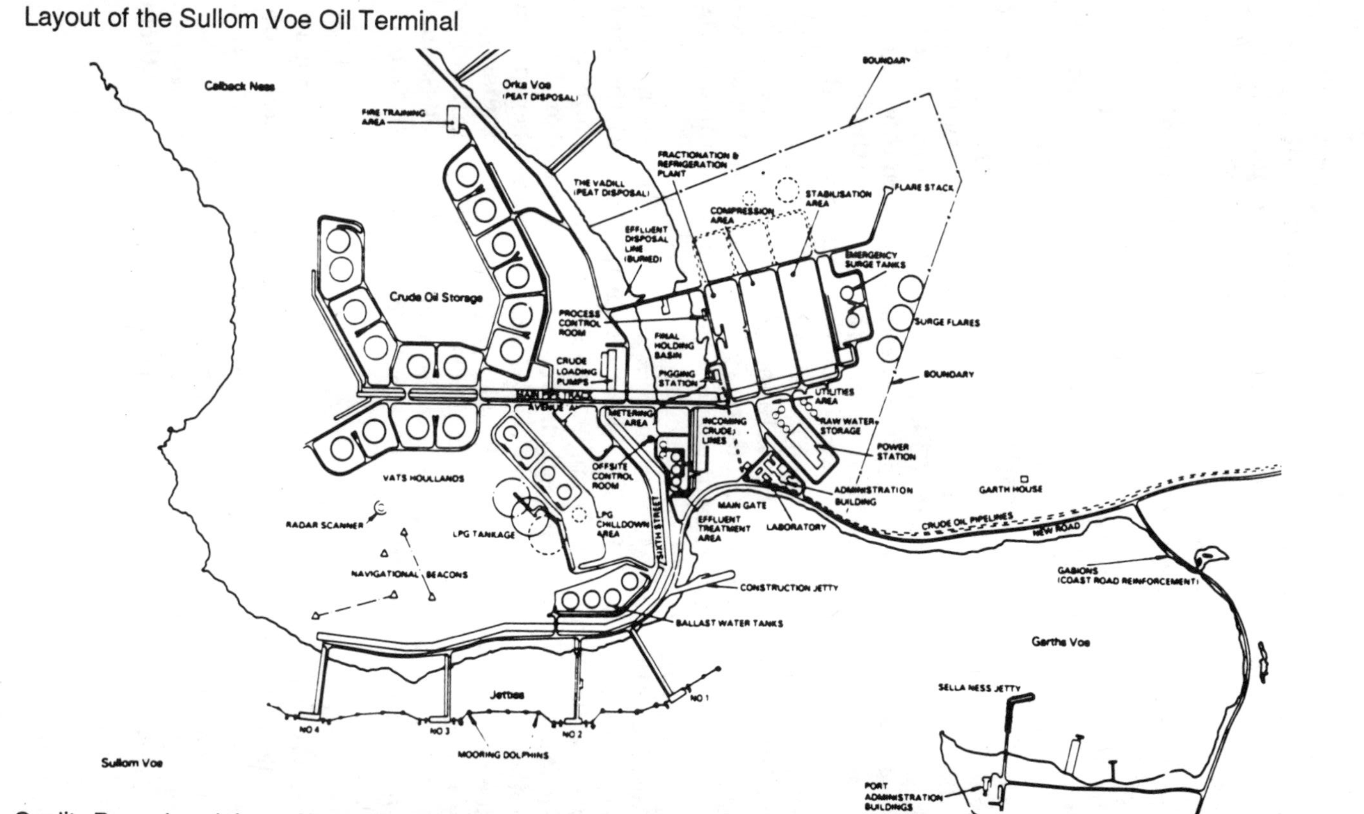

Credit: Reproduced from *Shetland's Oil Era* by Elizabeth Marshall with the permission of the publishers, The Research and Development Department of Shetland Islands Council.

Sullom Voe: how it works—Oil and Gas Flow Chart

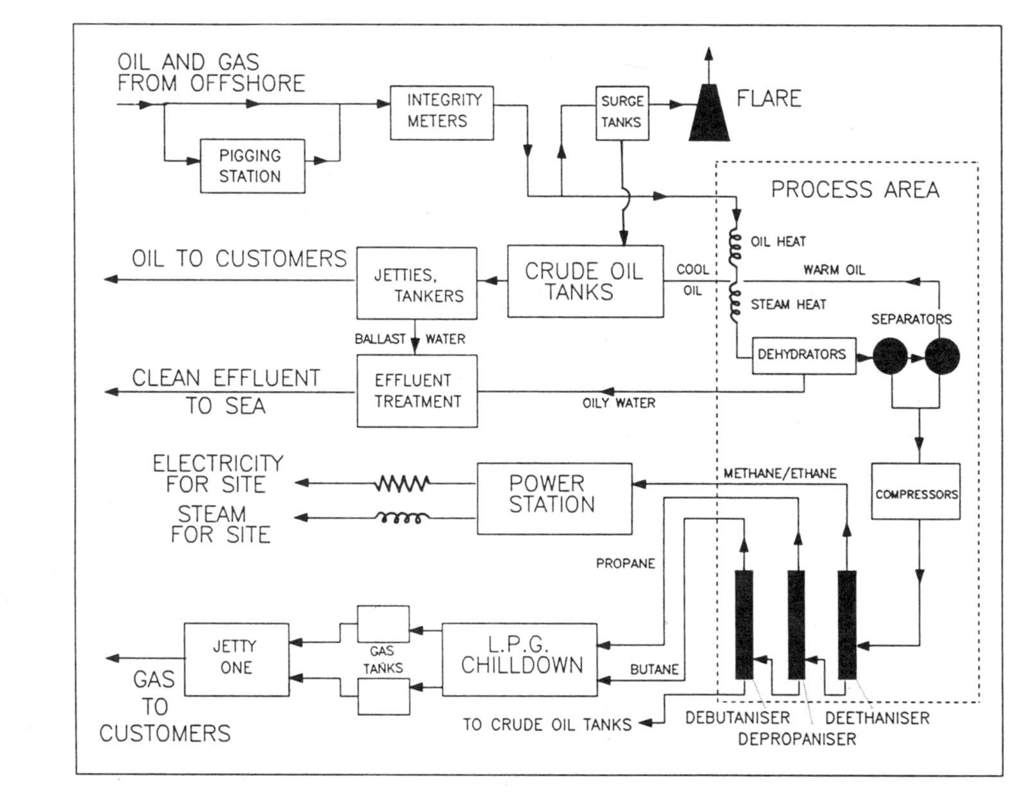

Credit: Erik Arthur and Jonathan Wills

The jetties were certainly a source of revenue, as we shall see when we come to discuss the finances of the council's oil era, but perhaps this revenue could have been gained in other ways. And finding the money to build them was to cause a serious row between the oilmen and the councillors. Only the accountants fully understand the jetty finance problem but, briefly, the idea was that the council would borrow the cash to build them and be repaid by the industry at 2 percent per year of this capital cost. The trouble was that no-one thought to put in a fool-proof clause in the deal, to cover the whole cost of borrowing the money as well as covering the repayment of the principal sum. So, when the council made some serious foreign exchange losses on several huge chunks of the loan, they were left with an unexpected bill. It was all sorted out, we are told, in the Busta House Agreement signed in November 1988.

Never mind, the jetties were built—at a cost of some £47 million at 1981 prices. This compared with the £1,300 million for the oil terminal, but it was still big league for the islands council. And then there was the cash to build the tug and pilot boat harbour on the other side of the water at Sella Ness, plus the port administration building, workshops and equipment stores, a high-technology navigation system, the pilot boats and the tugs for Shetland Towage, the council's joint venture company. There were those who speculated that it might have been cheaper to have followed Ian Cadwell's advice in 1972 and let the industry build the lot.

Building all this was a much more straightforward job than building the terminal, but it still rated as a massive civil engineering project. The water in Sullom Voe is deep, which meant that while the jetties could be fairly short, huge costs were involved in piling the tubular steel legs securely into the bedrock. And each jetty had to be equipped with enormous loading arms (whose leaking oil seals were a persistent cause of trouble), plus elaborate fire precautions involving gigantic fire pumps and, of course, the pipelines to load oil and gas and to take dirty ballast water ashore.

The story did not end there; the entrance channels in Yell Sound and the mouth of Sullom Voe itself had to be provided with automatic lighthouses, installed and serviced by helicopter. This was to be one of the biggest projects ever undertaken by the Northern Lighthouse Board based in Edinburgh. And the hydrographic division of the Royal Navy had to send survey ships for two seasons to revise the charts of the area, which had never before seen ships up to a 1,000 feet long and 90 feet deep. Some alarming and uncharted submarine rock pinnacles were discovered offshore.

Some day a more technical pen than mine will tell the full story of the building of the port and the terminal, but the important and remarkable point is that somehow it all came together in time for the oil revenues to start flowing into the coffers of the beleaguered Labour Government in the spring of 1979. Not that it helped them much. The irony was that the Labour ministers who had done so much to assist the passage of the special Act of Parliament for the Shetland council were to spend the next decade in the political wilderness. It was the Tories who reaped the benefit of North Sea oil and squandered it on the highest unemployment since the 1930s. In the process they created a society in which the sort of public enterprise displayed by Shetland Islands Council became anathema in Britain as a whole—for a while.

Those of us who had worried so much about the advent of oil and had raised all those environmental concerns in 1972 to 1976 had to eat many of our words, if not our hats. Astonishing technical feats were performed at Sullom Voe: it was deeply impressive, both in its construction and in its operation. It would be churlish to deny that, and unfair to the thousands of hard-working folk who knew and cared little for council/oil industry politics but just got on with the job. For all the managerial foul-ups, the beast was built. It remains a marvel.

As soon as construction work began to wind down, many of those who had worried out loud about the impact of oil in the 1970s began to worry out loud about the impact of losing oil in the 1990s. They need not have done; oil will be a fact at Sullom Voe well into the 21st century. Throughput is planned to decline to a plateau of around 250,000 tonnes a week (a quarter of peak production) but, if the more optimistic forecasts of the UK Offshore Operators' Association are correct, there is still a great deal of oil and gas to be had out of the seabed around Shetland. Even if a quarter of it can be attracted to the pipes into Sullom Voe rather than loaded direct into tankers from the wellheads, a chancy business in the winter, then Sullom Voe will be in business for another 25 to 50 years.

Even before the oilfields dry up, there will be some wells and platforms which come to the end of their useful lives, and parts of the Sullom Voe terminal which will have to be decommissioned and/or replaced. The process has already started, with the recent amalgamation of the two gas processing trains there. This will mean another job to do—taking all this ironmongery to pieces again. Scrap metal could well be the growth industry of Shetland in the 2020s. Assuming, that is, that the oil companies are held to their promise to reinstate the Calback Ness site and remove the offshore structures and pipelines. Somehow, I suspect that they will find a way around that problem too. Promises, promises ... that is the way it has always been, and that is the way it is likely to go on.

The site in October 1981: all 16 crude oil storage tanks are in use, but parts of the terminal are still being built.
Credit: BP

"A Disaster Waiting to Happen" 5

Sullom Voe, Captain Chris Hunter told the local press early in 1978, just before he handed in his resignation, was "a disaster waiting to happen". And happen it did. At 11.38 pm on 30th December, 1978, the oil tanker *Esso Bernicia*, 190,600 tonnes deadweight, collided with concrete mooring dolphins at Jetty #2. Out of the 25 feet long gash in her side spilled an estimated 1,174 tonnes of oil. She was the twelfth and largest tanker to berth at the port since it had opened for the first cargo a month before. It was not a large spill by the standards of the *Amoco Cadiz* in Brittany the previous winter nor the *Torrey Canyon* off Cornwall in 1967, but it was a very unusual and significant one. Nothing would ever be quite the same again for the oil industry in Shetland.

At the end of November there had been a great public relations performance from BP about "First Oil Ashore"—meaning their achievement of pumping into Sullom Voe terminal the first globs of sticky glub to come bubbling out of the huge pipelines from the oilfields 100 miles offshore. Now it was ashore all right, but not quite in the sense that the publicists had imagined. The irony was that the oil which polluted Sullom Voe had nothing to do with the terminal and its pipelines—it was heavy *fuel* oil from the *Esso Bernicia's* bunkers, the very worst kind of all when you need to clean it up in a hurry. Try cleaning a piece of carpet with a toothbrush and a teaspoon after you have spilt molasses on it and you will have some idea of what it was like. Even more ironical, the spill could not, by any stretch of the imagination, be said to be BP's fault, although the subsequent escape of the oil into Yell Sound, causing the deaths of more than 3,000 seabirds, was very much the fault of BP and the manufacturers of a defective floating oil boom.

That Sullom Voe and the Shetland coast in general have been relatively free from serious pollution since 1979, owes much to the measures which BP and their associates were forced to take in the oily wake of the *Esso Bernicia*, forced not by law, nor even by Shetland Islands Council, but by the urgent need to do something about appallingly bad publicity.

By the time the chain reaction of events which led to the Esso *Bernicia* disaster had happened, the Captain Hunter I mentioned earlier was far away, as was his boss, Captain George Biro. Captain Hunter was the council's first pollution control officer at Sullom Voe, and he was less than pleased with the arrangements for handling the big ships and the oil they would inevitably spill. After his views became public knowledge in Shetland, he had left for a new job in Wales, six months before Captain Biro's rather more spectacular departure in December 1978, to run an even bigger oil port for *real* Arabs in the Persian Gulf.

George Biro, a Hungarian by birth, was on first acquaintance reserved and correct to the point of brusqueness; a businesslike master mariner with a short fuse who stood no nonsense, said what he thought and kept his distance from the lower deck. Off duty, once he was sure of you, he turned out to be an entertaining companion with a hilarious line in reminiscences and a wickedly accurate grasp of his own and others' frailties. His wife Nan was immensely proud of him. Her pointed and irreverent remarks, laced with sarcastic fun, helped to upset Captain Biro's political masters in the tiny social whirlpool of Shetland.

Captain George Biro—the first director of ports and harbours for Shetland Islands Council and the man who planned "the Rolls Royce of ports".
Photo: J. Wills Archive (photographer unknown)

For the first Director of Ports and Harbours to be hired by Shetland Islands Council, it was all very simple or should have been. In 1975 Biro was asked to use his professional expertise, based on a lifetime at sea, to advise the council how they should set up and run the port of Sullom Voe. He helped to work out which channels the tankers could use, the navigation lights and radars they would need, the numbers of pilot boats and mooring boats required, and the number of tugs that would be necessary to handle ships of up to 500,000 tonnes in some of the windiest tidal water in the world. Most significantly of all, he insisted on salaried marine pilots employed directly by the council.

Many harbour pilots are self-employed, independent contractors providing services to the port authorities or ship-owners who hire them. Biro favoured a different system which he had seen working successfully in other large tanker ports around the world: the pilots would be employees, not contractors, and they would work shifts in the port control room and offices, as well as bringing tankers in and out. They would be called marine officers, not just pilots, and the system would ensure that the man in overall control of a berthing, watching the radars and listening to the radios, would know exactly what his colleague on the bridge of the tanker was thinking and doing.

From time to time the more rebellious marine officers were to demand, always unsuccessfully, what they saw as the more remunerative self-employed status. One of the ringleaders in this movement was subsequently given early retirement after errors in navigating a loaded tanker in January 1987. His deviation from the intended course was detected by fellow officers in port control who alerted the tanker's master in time to miss the island of Lamba.

The first battle between George Biro and the councillors was over what the marine officers should be paid. Their 1978 salaries of over £22,000 a year sounded gigantic to local people but in truth were rather less than what were available deep sea or even in other large ports. Biro argued, and many agreed with him, that if the port was to have officers well enough qualified to keep it clean and safe then the council would have to pay the price. It did not go unnoticed that, under Biro's proposals, a marine officer could end up earning more than the then chief executive of Shetland Islands Council. Not, as more than one local wag was to remark, that you would want your average chief executive steering 200,000 tonnes of crude between sharp rocks at the bottom of your garden.

Relations between Ian Clark, the chief executive, and his director of ports and harbours were already less than cordial when the next row flared up. Biro's considered opinion was that no fewer than six tugs were needed—to be sure of always having four tugs available to handle the larger tankers, plus one on standby and one away for routine overhaul. Not only that but some of the tugs should be of the "tractor" type—highly manoeuvrable vessels with *Voith-Schneider* propulsion units that resembled food blenders mounted on vertical shafts underneath the hull, unlike the conventional screw tugs with propellers on horizontal shafts sticking out at the stern. The argument was technical and complicated but basically Clark and his successor Ernest Urquhart, who took over in 1976, considered that three conventional tugs were enough, at least to begin with, plus a fourth in reserve. And that is what they got. Only on the fateful night of 30th/31st December 1978, the crew of the fourth tug were on New Year leave, "on call at home", as the subsequent council inquiry was to put it.

To be fair, both Clark and Urquhart were under considerable pressure from the oil companies to minimise the running costs of Sullom Voe, and not everyone in the council's joint venture tug company with the Clyde Cory company, Shetland Towage Ltd, agreed with Biro. The oil industry's position was made clear at a meeting in February 1978—they thought four tugs would be enough and that ships the size of the *Esso Bernicia* could safely be handled, depending on weather conditions, with only three. The fight had in fact started two years earlier when Clark had, according to Biro, told him, in so many words, to mind his own business when it came to spending money on tugs.

Clark resigned in 1976 to join Britoil, the short-lived state oil giant. His successor was a trained quantity surveyor. The advertisements for the post described it as "Britain's top job", a geographical pun which caused some hilarity in local circles. The successful candidate, Ernest Urquhart, was everything that Ian Clark was not: for starters, he was a Roman Catholic rather than a Plymouth Brethren, later Baptist, lay preacher and, although religious affiliation played no part in his selection by the Protestant and agnostic councillors on the interview board, he was a man less likely to wear his moral convictions on his sleeve and much less likely to try lay preaching in the council chamber.

Urquhart also lacked the physical presence of the tall, prematurely white-haired, thick-set Clark—he was a small, dark, dapper man in immaculate suits and with a facial expression that can only be described as sharp. In his grasp of the bewildering details of a rapidly expanding council, which now had fingers in nearly every Shetland pie, he mesmerised some councillors whose brains worked more slowly. And, when he did make mistakes, his verbal dexterity almost always got him out of the mess.

Clark had enjoyed an easy and relaxed relationship with the media. He was a natural on radio and TV and a prolific source of colourful quotes for newspaper reporters. Many of his pronouncements had a strong moral tone, and, despite his imposing personal appearance, he succeeded in encouraging the image of Shetland as David and the wicked oilmen as a collective Goliath. He was always accessible to journalists and willing to say *something*. He played the media along, and he knew exactly how and when to 'leak' information.

In contrast, Urquhart was a public relations disaster. Long years in industrial middle management had lumbered him with a mechanistic vocabulary in which intransitive verbs did amazingly transitive things—matters were "progressed", for example, or "actioned". There was no end of "input" and "output", just as in the process diagrams in the petrochemical works with which he was so familiar. Little or no information could be released to the public through the press and local radio until it had been sifted, sanitized and ratified by plenary sessions of the council. These meetings and the subsequent "press releases" might take place weeks after the event in question. Poor Mr Urquhart, highly trained and mentally agile though he was, had not the slightest idea how to put the message across. His utterly rational mind appeared to work like a flow chart, always in straight logical trains of thought leading to clear 'Yes' or 'No' answers. This was a characteristic which soon irritated his senior colleagues and the less over-awed councillors who in theory were running him. However, even more so than the charismatic Mr Clark, he was running them.

There was never any way chance that conflict could be avoided between the bluff Captain Biro, no politician on his own admission, and Urquhart, veteran of political infighting in the career struggles of big corporations. The clash, when it came, was in the most public of places. During a routine interview on the local radio station where Urquhart was asked about progress and problems at Sullom Voe, he suddenly launched into what Biro took to be an assault on his personal integrity, questioning his facts and, worse, his opinions about the tugs issue.

Biro kept his dignity, absurdly so, it seemed to many who sympathised with him, and refused to comment to reporters about this public dressing down. Shortly afterwards, following acrimonious exchanges behind the scenes, he resigned in disgust at his treatment from Urquhart and from the council's septuagenarian political boss, Convener A.I. Tulloch, who backed Urquhart rather than the ports director. He would probably have gone anyway, but this made it sooner rather than later.

It was a sad and typically wet and windy December evening when the Biros boarded the *St Clair* ferry to Aberdeen for the last time. Hardly anyone came to see them off. Those who did sensed that George Biro would have liked to stay long enough to see 'his' port reaching its designed capacity and proving his ideas in practice. He could not have guessed that, within a month, much of what he had said about tugs would become accepted by the council and everyone else in Shetland, nor that Captain Hunter's dire prediction would come true, with the most paradoxical results.

Sullom Voe was about to become the cleanest oil port in the world, as was later claimed, but first it had to become one of the dirtiest. It happened like this:

Esso Bernicia was a mucky-looking steam turbine tanker, 10 years old when her owners realised that she was the ideal size for the early cargoes out of Sullom Voe, just then coming on stream. So she headed for the Ramna Stacks rocks off the north Mainland of Shetland, where her black hull loomed out of the remains of a gale of wet snow in the murky evening of 30th December 1978. She picked up her pilot and steamed southwards down the west Yell Sound channel, past the sheep grazing seaweed on beaches below the snow covered crofts. In the wake of the 1,062 foot long ship were unseen, scattered flocks of Black Guillemots and Great Northern Divers, wintering on this bleak stretch of water as their predecessors had done for thousands of years and feeding on the rich marine life stirred up by the tides. Within a week most of them would be dead, suffocated in fuel oil or poisoned by it when they tried to preen their polluted feathers. Some of the sheep would be dead too, and the fleeces of many others ruined by tarry oil.

The weather forecast was bad—gusts of up to 30 knots in snow showers—but the ship had arrived during a lull to collect her cargo of oil for Slagen in Norway. It was calm enough at the pilot station, 12 miles north of the jetties, for Captain Tom Hemingway, the pilot, and his observer, Captain Bob Manson, to get aboard the ship from the pilot cutter. Two days before, the ship's agents had phoned an order for four tugs. The written confirmation asked only for three. This led to a row between Captain Don McElvogue, the duty officer at port control on the morning of 30th December, and his boss, acting ports director Captain Bert Flett. Captain McElvogue who, to prevent embarrassment to his superiors, was not asked to give evidence to the subsequent council inquiry, made a retrospective entry in the port control log after the accident, recording his version of events—four tugs had been asked for and refused by Captain Flett. By this time the crew of the fourth tug were enjoying their New Year holiday—at home but "on call", provided they could reach the terminal through the snow.

At first, everything went like clockwork. The tugs came alongside as the ship passed Little Holm, and two of them put lines aboard as the tanker came in past the isle of Lamba. After entering the harbour the *Esso Bernicia* was swung around facing the way she had come, and a line was put ashore from the ship's starboard bow to Jetty #2.

The south-easterly wind, gusting to 18 knots on her port quarter, was pushing her stern in towards the mooring dolphins. Counteracting this force was the tug *Stanechakker*, pulling at the *Esso Bernicia's* stern.

Down in the engine room of the *Stanechakker*, a faulty pipe coupling was under intense pressure as the tug used her hydraulic deck winches to take the strain of the 12 inch diameter rope from the tanker. Unseen, the pipe started leaking hydraulic fluid onto the hot exhaust manifold of her starboard engine. At 11:32 pm the *Stanechakker's* skipper radioed that his engine room was on fire. A minute later the *Esso Bernicia* got a stern line ashore. A minute after that the pilot and the master of the tanker gave *Stanechakker* permission to drop her tow and pull clear to put out the fire. They called in the midships tug, *Swaabie*, whose job that night was to push rather than pull and which therefore had no line aboard the *Esso Bernicia*. The plan was for her to get in under the stern of the big ship and take the *Stanechakker's* place. But it was too late. The crippled *Stanechakker's* winches were out of action. Her crew frantically hauled in 200 feet of heavy rope by hand but by the time they could get it out of the way of the relief tug's propellers, the tanker had slewed round. As the force on the stern line from the *Esso Bernicia* went beyond its designed strength of 125 tonnes, it snapped. For some reason never established, no one pulled in the bow line from the tanker to the shore to act as a "spring" and reduce the force of the impact. And, because the *Esso Bernicia's* engines were steam turbines, they did not respond quickly enough to the call for power when the tug was lost.

The helpless ship hit first one reinforced concrete mooring dolphin, then another and then another. There were no fenders on the sharp corners of the dolphins and the tanker's plates were burst like tinfoil. By five minutes to midnight, the combined efforts of her crew, the remaining tugs and the jetty crews had got the *Esso Bernicia* secured. The fire in the *Stanechakker's* engine room was out. On board the tanker the pumps were trying to empty the 3,000 tonne starboard fuel tank. But 1,174 tonnes of bunker 'C' were already in the water and drifting north along the shore.

How much of the oil went straight out of Sullom Voe and into Yell Sound is unknown. The relatively light south-easterly wind tended to push it onto the shore below the oil terminal. How much would have escaped if proper booms had been available to corral the oil immediately is not known either. What is known is that the spill was followed by a series of spectacular and farcical foul-ups. It took 11 hours to get the first boom into the water. The council's newly-appointed oil pollution control officer, Mr Alastair Grubb, had only been in Shetland for two weeks and did not yet have a phone in his house. Fortunately, his deputy lived next door, and the message that something was up was quickly passed on.

Within an hour of the spill, Mr Grubb and his men were at the scene, only to find that no-one had the key for a store where essential equipment was kept. The portable booms were covered in ice and snow which had to be chipped off before they could be launched. Other equipment was missing from an unsupervised store. Then they had to wait three hours until the driver of a mobile crane could be found to launch the Vikoma Seapack boom. In despair, they decided to leave it until daylight, when a helicopter was called in to find out exactly where the pollution was.

Two hours after daybreak, following a couple of trial starts, they tried to start the compressor which pumped air into the inflatable boom. Its battery was flat. Even so, they did get it into the water before lunchtime, and it did contain some of the oil. So did a second Seapack boom flown in from Orkney. On the 3rd and 4th of January both booms failed—one because of a faulty clutch which had been known about before, and the other because of a bearing starved of lubricating oil. The inflated sections collapsed and the oil flooded into Yell Sound, where tides and winds carried it far beyond. It got as far north as Unst and as far south as Whalsay. The organisational foul-ups bore an uncanny resemblance to what was to happen in Prince William Sound, Alaska, 10 years and 3 months later.

Meanwhile it had become depressingly clear, to those who had managed to get to Sullom Voe through blizzards which blocked many roads, that the only successful way to clean up heavy fuel oil was to shovel it into buckets and trashcan bags, by hand. Spraying dispersants was tried but, in snow showers with temperatures around freezing, the chemicals had no effect. The stuff was far too sticky to pump. So for six months, hundreds of men wearing oily oilskins laboured with buckets, shovels and bin bags to collect what they could from the beaches. It was a bitter, filthy task, and on the rockier shores it was hopeless. Eight years after the spill, traces of tarballs from the *Esso Bernicia* could still be found along much of the north and east coast of Shetland.

The oilspill made minor headlines for a day or two in the national media and was then forgotten by the outside world. It was 12 weeks before the national newsdesks woke up to the scale of the disaster, after repeatedly ignoring stories from local correspondents whom they perhaps suspected of crying "Wolf!" once too often. It took a photograph to do it—Gibby Irvine's *Shetland Times* picture of March

1979 showed a flock of bedraggled, oily black sheep with just one clean white one in the middle. By then, more than 2,000 sheep had been polluted as they went about their traditional winter grazings along the shore. Fifty of them had died. The sheep in the picture were from the island of Yell, across the sound from Sullom Voe. They had been rounded up to keep them from foraging among the seaweed which gives Shetland lamb and mutton its distinctive flavour.

The council was faced with bills for miles of new fencing to keep the flocks from the shore, plus strident demands for free hay and concentrate fodder for the animals. The bills were paid, slowly, first by the council and then by the oil companies and, "without prejudice [to the subsequent litigation to recover the money]", by the *Esso Bernicia's* insurers. Some cynics suggested, and not without evidence, that a few of the less scrupulous crofters, seeing a good thing in the making, had led their sheep at night to places where they could be sure of getting them oiled and thus eligible for compensation. There were also apocryphal tales of oiled sheep being trucked around from one croft to the next, keeping ahead of the inspectors who were sent to check how much damage had been done and where.

The clean-up cost, at a conservative estimate, was £2.25 million, plus over £300,000 in compensation to crofters for their oiled sheep and lost shore grazings and more than £250,000 in repairing Jetty #2, which was out of action for months. To have repaired *Esso Bernicia* would have cost more than she was worth—from Sullom Voe she went to the breakers' yard. At the time of writing, 11 years after the disaster, the arguments about who was to blame and who should pay the total bill of over £4 million are still rumbling through the halls of the Court of Session in Edinburgh. Was it the council's fault for not having enough tugs? Or was it the fault of the designer of the *Stanechakker* for putting a hydraulic coupling anywhere near a hot engine manifold? Or the builders, for not putting the tug together properly? Or Shetland Towage Ltd, for accepting a faulty tug? Or the way in which the tanker and tugs were handled? Or BP, for having a pollution control team and equipment that were totally unprepared? Or the people who made the faulty boom engines? All we know so far is that the council has been exonerated by the court and that the lawyers' fees will probably exceed the actual costs of the *Esso Bernicia* disaster if the case drags on for much longer.

Some costs could not be ascertained. No-one claimed compensation for the deaths of 146 Great Northern Divers (Loons), 683 Shags, 306 Long-Tailed Ducks (Old Squaw), 570 Eider Ducks, 336 Guillemots, 633 Black Guillemots and 14 Otters. More than 90 percent of the seabirds in Sullom Voe and 75 percent in Yell Sound were killed. The commoner species, such as the Eider Ducks, which live in Shetland all year round, soon recovered and re-colonised the polluted areas. Others, such as the Loons, wintering in Shetland from Eastern Canada and Greenland, did not re-appear for years and then in smaller numbers than before. These birds are long lived, conservative-minded creatures, and the same individuals had almost certainly been coming to their favourite spots in Sullom Voe and Yell Sound for many years. Now they were dead and, with only about 300 or so Loons in Shetland during the winter, it may be many generations until their successors discover the empty *niches* and fill them.

The oil from the *Esso Bernicia* was not the only cause of the shocking bird mortality around the islands that winter. To the death toll of 3,704 birds of 50 different species were added at least another 2,000 killed by unscrupulous tanker masters, swilling out their dirty bilges as they approached the coast of Shetland. This trick saved them time, and therefore money, because it meant they did not have to pump dirty bilges ashore for treatment at Sullom Voe before loading their cargoes. Word of the *Esso Bernicia* spread quickly, and the temptation to dump dirty water in the sea, in the hope that it would be blamed on the unfortunate Esso tanker, was too much for some of Sullom Voe's customers and other passing ships. Another 2,000 seabirds were killed around Orkney at the same time and from the same cause.

The Shetland Times summed up popular feeling about *Esso Bernicia* in an editorial comment when the council's report on the disaster was eventually published in June 1981, minus the actual evidence, all of which the inquiry committee had taken in secret. The editorial demanded that the evidence should be published. The inquiry report made "sorry reading" and the delay in publication was "lamentable". It was clear that Sullom Voe should never have been opened in November 1978 when "on the pollution control side it simply was not ready for business". The paper brusquely dismissed the excuses: "Saying the weather was bad is no excuse—the weather is often bad..."

Most damaging of all was this sentence: "It was opened before it was ready, for one simple and rather unpleasant reason—the oil industry's need for revenue was allowed to outweigh the sensible environmental safeguards that Shetland was entitled to then and has since, belatedly, acquired". The editor might have added that the council's own need for revenue, from the penny a ton levy on all oil shipped and from tug charges, pilotage dues and other exactions, was also somewhat pressing.

It was not just the oil industry that had been caught out. The general suspicion was that the council had acquiesced in an avoidable catastrophe.

It was not the case, as was later claimed by Urquhart and others, that the council had no idea of what might happen if oil were spilled. In April 1978, three weeks after the *Amoco Cadiz* was wrecked off Portsall in North Finisterre, Captain Hunter and Mr Billy Smith, the council's director of construction, visited Brittany at the council's request to see for themselves how the French authorities had coped, or rather, as the two men were to point out, not coped, with the pollution disaster. Their report was available to the Shetland Islands Council on 9th May 1978, more than seven months before the *Esso Bernicia* disaster. Little immediate action appears to have been taken on this report. It still makes interesting reading.

Mr Smith and Captain Hunter were perhaps thought to be having a little joke when, in a comment on advertisements taken out in the French press by Breton farmers to assure the public that their produce was unaffected by the pollution, they told Mr Urquhart and the councillors: "One can foresee what effect a report in a British newspaper of Shetland sheep eating oil contaminated seaweed would have on the sale of Shetland lamb".

They went on to criticise the inflatable booms used on the Breton coast and noted that there did not seem to be any equipment available for the physical recovery of the oil—shovels and buckets were the most commonly used recovery methods, often wielded by volunteers. They also pointed out that migratory birds, especially the Loons, had been "severely affected".

In conclusion, the Hunter and Smith report warned councillors of the lessons from *Amoco Cadiz*: "The French authorities ... were obviously unprepared for a disaster of this magnitude, though in fairness one must state that nor is anyone else. ... Booms, when eventually available, did not for the most part work. This was due partly to the lack of preparation and partly to the prevailing winds.

> The apparent absence of mechanical devices for the recovery of oil ... spraying equipment and [ab]sorbent material was remarkable but not as remarkable as the statement given in a television interview by the Government minister responsible for the clearing of oil pollution that Plan Polmar [the French authorities' emergency oilspill plan] was working well. Most Bretons were obviously in disagreement with this view and it must be said that they had ample evidence to support their case.

After the Shetland disaster, the criticism of our state of preparedness (or the lack of it) was by no means confined to the local press. In 1981, in an editorial comment in the prestigious *Marine Pollution Bulletin*, the respected, normally restrained, ornithologist Dr Bill Bourne was even more scathing than *The Shetland Times* had been about the council's long-delayed report. Dr Bourne was a force to be reckoned with—he knew Shetland well and, unlike the journalists on the local paper and radio station, had been closely involved in scientific monitoring work connected with the Shetland Oil Terminal Environmental Advisory Group (SOTEAG).

He noted, unkindly but accurately, that the report had been released "under cover" of the "furore" surrounding the small IRA bomb which exploded in the Sullom Voe power station during the Queen's visit to "inaugurate" the terminal in May 1981. He said the council's ultimate conclusion was that the inquiry "would have been much better handled by an outside body with better technical assistance and a more experienced, legally-qualified, chairman.... .

"Presumably owing to legal considerations", Dr Bourne continued, "they appear either to have failed to secure sight of most of the more critical relevant technical reports on what went wrong or, if they did, fail to quote them.

"It is quite remarkable", he sneered, "that the only individual who is found in any way to blame is the person [Captain Don McElvogue] who dared retrospectively to insert four words revealing the basic cause for the accident [i.e. the lack of a fourth tug] in the port control room communications log...when he was not on duty at the time so that it was none of his business."

The report left "all the more interesting questions unanswered", he concluded, but it was "unreasonable to expect total candour from all concerned". The true culprits were the Government ministers and officials who had allowed Sullom Voe to be opened when the council was "insufficiently experienced to be left in charge of a major oil port". The official verdict that it was "just bad luck" was not good enough, Dr Bourne wrote. "Something more positive" must have been required to produce "such a shambles".

Such criticisms deeply embarrassed the council, particularly when people of Dr Bourne's standing rubbed in the fact that the disaster had cost the council hardly anything. By contrast, the oil industry had been "less lucky", he said, and had had to bear "their larger share of the surely excessive cost as well as the odium of the disaster".

Dr Bourne was to some extent wasting his ink. There was not going to be a proper inquiry until the excessively delayed one still going through the courts. Both the council and the oil industry were trusting to luck that eventually the odium of the debacle would fade and it would in time come to be seen as a blessing in disguise, on account of it being the stimulus to introduce pollution safeguards at Sullom Voe which set new world standards. This strategy worked, and, by the time the council report was published, most of the new regulations were in place. There were no more large oil spills, the local paper was full of stories about tanker skippers being fined astronomical sums for spills of a few hundred gallons and only cynics were still asking why all this had not been done before 30th December 1978.

So what had been done? Immediately after the disaster there were the usual calls for new laws by publicity-conscious MPs, environmental groups and such councillors as had not jumped on Urquhart's whitewash wagon. New laws take time and there has been only minor progress since *Esso Bernicia* in tightening up international law which, of necessity, is the only way to do these things legally, for the oil trade and the marine environment are, by definition, trans-national phenomena. Where the council and the oil company partners in the Sullom Voe Association were able to recover some of their credibility, was in choosing a completely different route to the desired goal. The vehicle to take them there was the fine print in the commercial contracts into which every tanker must enter before entering Sullom Voe.

By the summer of 1979, all inward bound ships were obliged to make radio contact with Sullom Voe Port Control when they crossed a point 200 miles from the terminal. They had to report how much clean and dirty ballast they were carrying, what state their safety equipment was in, and their course, speed, position and estimated time of arrival at the pilot station. Tankers heading for Sullom Voe were required to keep at least 10 miles from the coast of Shetland, to pass midway between Shetland and Fair Isle or between Orkney and Fair Isle and on no account to navigate through the shoaled water between such islands as Foula and Out Skerries and the Mainland of Shetland. Several tankers had been observed steaming as close as a mile from the rock infested shores of Shetland in previous months.

On top of all that, they were not to be allowed to enter the port unless at least 35 percent of their cargo and ballast space was full of water—thus removing the financial incentive to dump dirty water ballast at sea in order to make a quicker and cheaper getaway from the loading jetties. The rules on closing the port in high winds and fog were also tightened up.

Why we worried—the Drongs rocks off the west coast of Shetland. A good reason for Shetland Islands Council's 10 mile "no-go" area for tankers.
Photo: Jonathan Wills

The sanction to enforce this was so simple that some wondered aloud why it had not been done before the terminal opened and, indeed, why the environmental advisory group, SOTEAG, had not suggested it. Any ship breaking the rules, set out in great detail in a telex sent to each and every inward bound tanker when she crossed the 200 mile line, would not be allowed to collect her cargo. She would in effect be banned until her owners could satisfy the port authorities that their ship would behave herself on future visits. A lost cargo meant a lot of money, as several irate shipowners and charterers were soon to discover.

An airplane (from 1989 an even more expensive helicopter) was hired by the council to make low level reconnaissance flights each day, inspecting ships trading to Sullom Voe and any other shipping suspected of making a mess. Any ship discovered to be the cause of pollution in Sullom Voe or around the Shetland coast while on passage to or from Sullom Voe could also expect not to be welcomed back, as a Greek tanker by the name of *Mihalis* was to discover. Her owners lost a court action in which they claimed that the council's spotter plane evidence of her dumping dirty ballast was insufficient.

All this and more was agreed to and paid for by the oil industry partners of the council in the Sullom Voe Association. There was local scepticism at first but this evaporated when it became clear that this time the oil companies meant exactly what they said. They rarely lifted a finger after 1979 to help a tanker in trouble with the port authority at Sullom Voe. The new system produced a dramatic improvement. The monthly surveys of Shetland's beaches by volunteer ornithologists began to turn up fewer oiled corpses. There were no more emotive pictures of suffocated birds in the press and on TV. There were still some oiled seabirds, of course, but they were mostly the result of spills from drilling rigs and oil production platforms far out to sea. It must also be admitted that the harbour at Lerwick is not exactly clean these days and usually a great deal oilier than Sullom Voe. But this pollution comes from fishing boats and factory ships which appear to be above the law in Shetland, for reasons of patriotic sentiment and parochial greed. "Over the side is over", say the Government's posters promoting the new anti-dumping laws. It is not over in Shetland.

The big oil disaster could still happen; but the new regulations meant that if a giant tanker, fully laden, were to break down on a lee shore there was a better chance than before that tugs would be able to get to her in time to prevent the ecological catastrophe of 250,000 tonnes of crude oil wiping out some of the finest seabird colonies left in Europe. If the pollution were to happen in the harbour of Sullom Voe, well, things were suddenly a lot better organised there, too. After a lengthy interval and much wrangling, seven properly-designed spur booms were installed at great expense at strategic sites around the voe. The council and the oil industry refused to countenance a large boom closing off the narrow entrance to Sullom Voe altogether. They said it wasn't feasible because of the currents. In fact there is hardly any current, and perhaps the reason was the potential loss of revenue to both council and oil companies if the harbour had to be closed every time there was an oil spill. But the spur booms were an improvement and did prevent the escape of oil on several subsequent occasions. New purpose-built anti-pollution boats were built, five tugs were provided—two of them the *Voith-Schneider* type which Captain Biro had specified; large quantities of specialist equipment were delivered; and, just as important as all the high-tech skimmers and oil recovery devices, a store was filled with buckets and shovels, just in case.

The council's oil pollution control team and the BP clean-up squad across the water at the terminal were re-trained and re-equipped to the standards which SOTEAG should have demanded in the first place. They even installed a computer which lists the detailed casualty history of every vessel arriving at Sullom Voe and

Bulbous bows in Yell Sound. One and a half inches of steel separate crude oil from salt water.
Photo: Jonathan Wills

can tell Port Control at the push of a button or two whether the ship has misbehaved at Sullom Voe at any time in the past. It is all very impressive. The present pollution control officer, Jim Dickson, delights in showing off his expensive and effective ironmongery to visitors. Unlike some of his colleagues, past and present, he also makes a point of ringing up the local media when accidents happen, giving the news through the front door before it gets out of the back door. It makes life for local journalists less conspiratorial, but they can always turn to BP across at the terminal itself where the PR people have still not got the hang of things.

One is left wondering how BP's PR outfit would cope with a huge oilspill inside Sullom Voe, where to this day the available equipment can only handle 2,000 tonnes of oil—a fraction of what even the Valdez terminal in Alaska was geared up for in March 1989. The BP answer will be that plenty more capability is on tap at 24 hours' notice, but the problem will be that 24 hours is a long time in the life of a big oilspill. Valdez is now looking at the two hour response to the big one.

BP's public relations certainly lost the place immediately after the *Esso Bernicia* disaster. One incident may illustrate how little they understood of the local public reaction to their "first oil ashore". In late January 1979, three weeks after the spill, the PR chaps invited press and broadcasters to dinner in the Queens Hotel in Lerwick, along with sundry local worthies and concerned environmentalists. I was invited along, being then in charge of the local radio station, BBC Radio Shetland, where we had taken a keen interest in the affair and had campaigned to have the full facts revealed.

On my way to the function, I met a friend from the Shetland Bird Club, largely composed of members of the Royal Society for the Protection of Birds (RSPB). He wondered if I would be good enough to hand round a press release they had prepared for the assembled media folk. Nothing simpler, I assured him, and innocently entered the Queen's Hotel, where I distributed it to my fellow hacks. I had only time for a quick glance at the document which was to cause such offence. It read as follows:

"The bird club is most concerned about the apparent inability of the council and the oil industry to deal with a small oil spill occurring in inshore waters in near-optimum conditions. The spill ... clearly demonstrated the shortcomings of both men and machinery and we have grave misgivings about the council's and the oil industry's ability to deal with any future spill in this biologically sensitive area ..."

The press release also had a go at Professor George Dunnet of Aberdeen University, the chairman of SOTEAG. He had been, to say the least, cautious in his comments and had carefully avoided condemning anyone. "Ninety five percent of the birds in Sullom Voe and 75 percent of the birds in Yell Sound at this season have now perished and we must ask Professor Dunnet whether he feels this constitutes an ecological disaster," the bird club said. "In the event of a negative reply, we invite him to define what does constitute such an event in the Shetland context."

The bird people complained about the apparent run-down in the clean-up operation and warned that hundreds more birds could die as long as hundreds of tonnes of oil were "unaccounted for". They concluded by demanding improvements in anti-pollution equipment, personnel and training, all of which were subsequently to be implemented.

Some of the oily people were a bit distant with me during the dinner but this was nothing new and it went off as pleasantly as possible in the circumstances. A few days later I received a letter from BP's PR chief in London.

"I was most surprised," he wrote on 24th January, "to hear that you took advantage of your position as our dinner guest ... to issue a press release on behalf of another organisation ....

"In not seeking our permission—which, incidentally, would undoubtedly have been granted—you not only caused the BP hosts to raise their eyebrows but also caused members of the press to comment adversely on your action. We always endeavour to keep a friendly atmosphere between ourselves and the media, but this unexpected action by you, I think you will agree, entitles me to bring our view to your attention."

Unlike the mainland UK reporters present, I had not had my fare paid to attend the dinner, and I confess to taking some pleasure in composing my reply.

"I must say that I'm surprised as well," I told the poor PR man, "surprised that BP's liaison with the RSPB, their fellow members of SOTEAG, is not good enough for them to know that the society proposed to circulate a press release. I had assumed that, like myself, you were good friends with the RSPB."

"As for the eyebrows of your local representatives, I can assure you that they remained entirely horizontal throughout the horrifying scenes you so graphically describe. I thought I was doing a favour for a mutual friend and I am sorry if it was not so."

I could not resist adding the suggestion that his office could spend "less time complaining about imagined breaches of etiquette" and more on English composition: "The press releases issued recently on behalf of yourselves and your hosts in Shetland, Shetland Islands Council, have been less than informative about the oil spill and not exactly written in clear radio copy", I concluded. Memos like this were to get me into trouble repeatedly during my time at the BBC, and this one almost certainly found its way into my personal file at the corporation where, knowing their secret-service style personnel department, it will remain for ever and a day.

That was 11 years ago. These days the anti-pollution arrangements at Sullom Voe satisfy both the bird club and the RSPB. The system is so impressive that in 1984 the environment committee of the European Parliament began consideration of a draft European Commission directive which would extend to every oil port in Europe the stringent requirements now in force at Sullom Voe. Knowing the European Parliament, we should see urgent action on this sometime in the year 2090.

By the way, I was not the only local resident to upset BP by uttering home truths while accepting their hospitality. Much more important people were also "at them". In that same January, Shetland's choleric and outspoken Lord Lieutenant, the late and much lamented Robert Bruce of Sandlodge, had his say.

Mr Bruce, some of the oily folk thought, abused their hospitality at the official opening ceremony by making pointed, not to say threatening, remarks about their performance at Sullom Voe since November. He warned them, in the plainest of language and in his capacity as the Queen's representative, that they had better pull their socks up. This public outburst, after he had downed a couple of pink gins paid for by his guests, was warmly applauded at many Shetland croft firesides where there was normally little time for landowners such as the old Lord Lieutenant.

In June 1979, five months after this outburst, BP decided that the time had come to lay on a rather special jaunt for the world's press. Now that Mr Bruce's remarks had faded in the public memory and there was actually some progress to show, BP desperately wanted to explain to the journalists what a splendid set-up they now had at Sullom Voe.

It was a memorable excursion. Dozens of energy correspondents, TV anchorpersons, financial journalists, radio reporters and news editors who knew a good trip when they saw one were flown to Shetland and joined the local paper and radio station in the Brae Hotel, a few miles from the terminal. After a merry day being bussed around all the pipes and tanks at the terminal and meeting white-coated supervisors who assured them that absolutely nothing could ever possibly go wrong again, there was a minor disaster when the hotel burnt the steaks. But more wine was quickly produced and few cared when or if the frazzled meat and two vegetables were served. PR men and women charmed their way through the throng at the bar, and it was with sore heads that the survivors rose at an early hour to get on board a tanker.

The BP ship *British Commerce* was more than welcoming. Reporters on the bridge observed her following every new rule in the book as she cast off from the jetty and headed for Rotterdam; there were tugs all over the place; the famous telex with the new "voyage instructions" was shown to them; the ship stayed at least 10 miles away from Foula and passed exactly midway through the Fair Isle channel; there was sunbathing on deck all afternoon; no-one was seasick, although a few passed out from over-indulgence; there was a vast and delightful meal, served by Goanese stewards—earning a ninth of what British crew would be paid, but no-one thought to ask about that; the reporters knew their manners, even if Mr Bruce did not—followed by liqueurs with the officers (all white and British).

It was early evening when the *British Commerce* hove to off the Scottish coast and transferred her precious cargo of scribes to a pilot boat for the journey to a hotel in Wick, where there would be even more free drink before the free flights home next morning.

This had all cost BP a fortune. It was well worth it, because not a single one of us swaying scribblers noticed that BP had just stopped a fully-laden tanker dead in the water, with no tugs or emergency cover of any kind, and for no reason except to offload a few drunks on a PR spree, a mere two miles upwind of a rocky shore festooned with thousands of seabirds. Had this incident been detected by the Shetland Islands Council's spotter plane and its spy in the sky video camera, the *British Commerce* would have been banned from Sullom Voe for ever. But it had not taken place off the Shetland coast. The ploy had worked. The resulting "copy" was favourable, on the whole. BP and their friends were once again seen trying, and mostly succeeding, to be Mr Kleen.

Parallels at Sixty North 6

It is no coincidence that several of the key people who built, commissioned and operated the Shetland oil terminal had also "done time" with BP in Alaska. The sixtieth degree of latitude is not the only parallel linking Sullom Voe and Valdez, 4,000 miles apart across the Arctic Ocean: the climate on the Gulf Coast of Alaska is similar to Shetland's, although even more robust. It also created similar difficulties in building, operating and maintaining the two terminals; and the winter storms brought identical problems in handling very large ships in constricted, rock-strewn waters of inestimable wildlife value.

In the early 1970s, BP undoubtedly learned many lessons in Alaska that they put to good use in Shetland very soon afterwards, not least in handling construction workers and unions on remote building sites, dealing with environmental concerns, and battling on through all that appalling weather. They also learned a good deal about manipulating local public opinion in Alaska.

What is interesting is the lesson they learned at Sullom Voe in 1978/79 but did not apply in Valdez until after the 1989 *Exxon Valdez* oil spill. In Shetland they boasted to reporters about the "Rolls Royce of ports", after their historic deal with the council in 1979 to protect the environment from another *Esso Bernicia* spill. In Alaska they let things meander on as they were, with even more catastrophic results than at Sullom Voe.

Granted *Exxon Valdez* was an Exxon problem. So what had BP to do with what went on at Valdez? They owned most of it, for a start. Certainly Exxon, as partners in Sullom Voe, had no excuse for ignorance of what went on there. But BP were not only partners; they were the builders and operators of Sullom Voe from 1975 onwards; they were also major shareholders in the Alyeska Pipeline

Service Company Inc., which ran the Valdez terminal on behalf of the seven participant companies in the trans-Alaska pipeline. By 1989 they owned just over 50 percent of the company; its boss was a BP man, one George M. Nelson; and a BP environmental scientist, Bryan Sage, who had worked for the company in both Alaska and Shetland, as early as 1975 had expressed his concern at the risks they might be taking. Needless to say, Mr Sage no longer works for the company, having opted for early retirement in 1978. He was not the only would-be whistleblower among the company men. Others had reservations about the corners that were cut during construction (leading to serious corrosion problems on the pipeline and in the terminal), about how the Valdez terminal was run, and about how it was affecting the environment of Prince William Sound—long before *Exxon Valdez* hit the Bligh Reef.

The horrible truth is that Sullom Voe was lucky to have its big spill right at the beginning. When Alaska's turn came at Easter 1989, and, let us not forget, it was 40 times as bad as *Esso Bernicia*, many of the malpractices that caused it were built into the system. *Exxon Valdez* was a catharsis. Since it happened, Valdez has caught up with Sullom Voe in some of its environmental safeguards—although Sullom Voe is now imposing even tighter safety standards. But it might have been avoided if BP, Exxon and partners—but especially BP—had put their money in Alaska where their mouths and their money were in Shetland.

Let us recall, briefly, the appalling events of the morning of 24th March, 1989. The first people on shore to see the *Exxon Valdez* in trouble were Jon and Susie Rush, two of the half-dozen remaining inhabitants of the ghost town of Ellamar, near Cordova, Alaska. Ellamar overlooks Bligh Island and the buoy marking the Bligh Reef, some 25 miles south west of the Valdez jetties. Jon earns a living as an artist—his delicate pen and ink drawings have made him a celebrated figure in south-central Alaska and further afield; he also does a bit of fishing in his superbly maintained boat, *Windsinger*. I met him a month after the disaster, at the Club Bar in the Prince William Sound port of Cordova (pop. 2,500), where he fetches his stores and has a scotch on the rocks or two with the boys. The Club Bar is 99 carat Yukon Trail, the decor of driftwood, fishing corks and old shipwreck beams redolent of another age.

"Is this part of the original Cordova, from the copper mining days before the war?" I asked him, over our third scotch on the rocks and the din of a live, electronic, rockabilly band from San Francisco.

"Hell, no! I designed and fitted out this place myself the fall before last!" Jon replied, squinting over his beard through bottleglass lenses. Now his subsistence fishing was in trouble because of the spill, and he had brought *Windsinger* across to Cordova to see if he could get hired by the clean-up contractors. I hated to think of that beautiful boat covered in oil.

I suppose I took to Jon Rush because he seemed to be an escapist hippy like I was when I went to live in Unst in 1973, only he had stuck it out and made it all come good. Now he was a successful middle-aged hippy, still doing his own thing. I was jealous. Anyway, I wanted him to tell me about what he had seen that night, after Captain Joe Hazelwood had discovered that his ship was aground on a rock.

"We were up late", Jon told me. "And round about one in the morning my old lady said, 'Hey, Jon, something's happening out there. All those lights, looks like a city'. In the morning when I saw it I thought it was inbound. I said, 'Susie, it's OK, he's inbound'. But he wasn't. The waterline showed that.

"To actually try to drive that thing off those rocks under his own power is beyond me. Was it total, blind panic? I don't blame the captain personally. They all fucked up. They should do it like the KGB, where they all watch each other."

I nodded. There did not seem much else to say so we had another scotch on the rocks. The band started to sound real good.

Ellamar was spared the oil. It blew away out to the west when the storm came up three days after the stranding. But within 10 days all the mussels and barnacles had fallen off Jon's pier, dead, apparently poisoned by the fumes from the stuff—what the experts call "aromatic hydrocarbons". The sea otters disappeared, although the Ellamar ones had not been oiled. He found three large squid dead on the shore, again without oil on them. But on the sonar from the bridge of *Windsinger* he could see dark masses of oil suspended beneath the surface when he went out to try to help the Cordova people with the clean-up. By that time Prince William Sound was in chaos.

The story has been picked over in microscopic detail by the media ever since. We all know what happened. The question is, how? From my investigations in Alaska and Sullom Voe, it is clear that it was not just a negligent skipper on, or rather not on, the bridge. Let us imagine it happening in Yell Sound, not Prince William Sound: the parallels are instructive—and destructive of BP's global posture as "Mr Kleen". The best way to tell the story is to go back over my notebook from the moment the airport bus dropped me at the corner of Main and Second in Cordova, Alaska, on the afternoon of 28th April 1989. I was 36 hours out of Lerwick, by my confused biological

Alaska's 500 Mile Spill

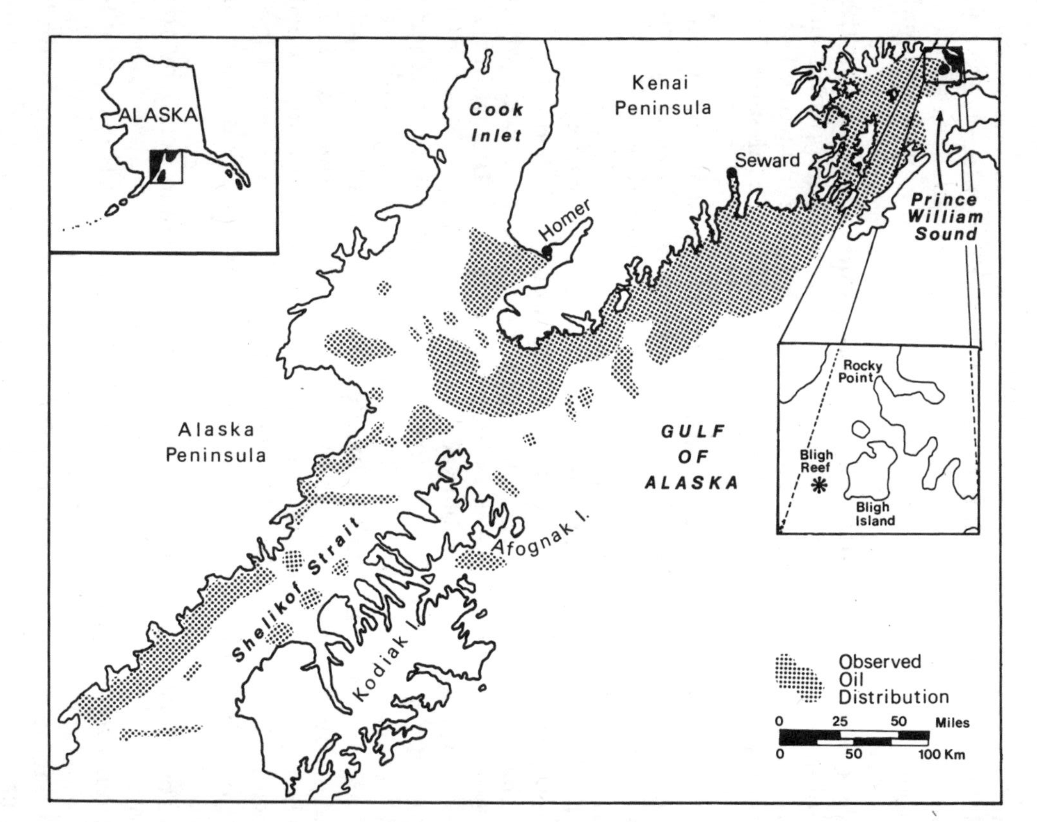

Cartography: Donald Battcock, from a map published by the State of Alaska.

clock, and supposedly writing a piece for *Condé Nast Traveler* magazine in New York.

Main Street, Cordova. Bits of this little fishing port look like a movie set for a re-make of Dangerous Dan McGrew. Jon Rush's Club Bar decor was right in character. There is even some real boardwalk outside the chiropodist's—left over from the footslogging days of the copper mining Klondyke of 1911. All we needed were some goodies to burst in the door of the bar and drill the baddies full of lead.

The baddies' office, in a garage up the hill, had carpet on the floor—Exxon's "Spill Response Center". The goodies were hanging out across the street from the Club Bar in the dowdy, chaotic union hall of Cordova District Fishermen United. In the back office was the man I had come to see, Jerry McCune, the fishermen's president. He was on the phone, with his head in his hands. Someone was telling him about the oiling of his favourite herring fishing ground. When he eventually looked up, I thought at first that he was drunk. But the hollow, red eyes were from four weeks without proper sleep. He was a mean looking man and right now he had a lot to feel mean about.

A long conversation began. It was to last for four days, punctuated by the endless phone calls and joined, as the hours rolled on, by dozens of other fishermen, office workers, salmon hatchery people, Greenies, marine biologists and their volunteer helpers.

I asked why the Valdez Coast Guard's vessel traffic system had not seen the tanker on radar when she failed to turn back into the channel after changing lanes. Bitter laughter—the guys on watch had been doing something else; they were not monitoring the screens; anyway, the Bligh Reef was at the limits of radar coverage. Commander Steve McCall, the US Coast Guard port captain at Valdez, later confirmed this to me in an interview.

"Come on", I said to Jerry McCune, "Surely it's like Sullom Voe; it's the same companies running it, after all. Surely they watch a tanker until she's past the last navigational hazard before the open sea? And a pilot must have been on the bridge, even if the skipper was drunk?" More bitter laughter. I must be joking; the pilot had left the ship just after she passed Rocky Point, about half an hour before she struck. There was no radar watch from shore and no pilot on board. The third mate, who was actually in charge on the bridge, did not have a pilot's licence for Prince William Sound. The lookout had seen the Bligh Reef light beacon and had raised the alarm. The ship had altered course, but too late to avoid hitting the reef and tearing holes in eight of her 11 cargo tanks.

When Valdez had first started up, I was informed, they used to pilot the ships out to the open sea at Hinchinbrook Entrance, and they used to have a more powerful radar, but there had been cutbacks. *Exxon Valdez* was on her own out there. Commander McCall was also to confirm this and his predecessor, Commander Jim Woodle, said the same when I asked him about normal procedures during his time at Valdez.

Back to Cordova; by this time we were tucking in at the Moose Club's crab boil night and I still had not had those four hours of sleep in the hotel that I had promised myself before we got down to details. OK, so the ship was on the reef, Captain Hazelwood had been called to the bridge ... why was he apparently allowed to try to drive her off that rock? From what we knew, it seemed that this could have caused a capsize, releasing all of her cargo into the sound rather than only a fifth of it. Jerry McCune and company could not answer that one, but a few days later the *Anchorage Daily News* did when they got hold of the vessel traffic system tapes—something that was not possible for *The Shetland Times* when the *Esso Bernicia* incident happened, but then this was America. The answer was that Commander McCall *advised* the skipper to take it easy and not do anything that might make matters worse. Advised, not ordered. But, I said, in the port limits of Sullom Voe, which extend out to the last navigational hazard, the port director could have ordered him not to do that but to wait for assistance. He could not do this in Prince William Sound.

I asked about the oil spill contingency plan. Which oil spill contingency plan, they said—there were eight plans drawn up by various agencies, resulting in total confusion. No-one was in overall control, equipment was missing or inadequate, and staff were poorly trained. A specialist team of pollution fighters had been disbanded some years previously. And then they told me about the incredible delays in getting booms, oil skimmers, pumps, lightering barges and the rest out to the hulk; how, even when another tanker was brought alongside to pump off the remaining cargo, oil kept leaking—the *Exxon Valdez's* pumps were used and a hose from them had holes in it; how no-one laid a boom around the stricken leviathan until it was too late; how the biggest industry in the world had let two days of perfect oilspill recovery weather go by, until the south-easterly gale came and blew it almost 500 miles from Bligh Reef; and how, in despair, the Prince William Sound fishers had taken their own boats out and started to do what they could—with nets, buckets, squeegees and anything they could lay hands on. They told me about their desperate efforts, eventually successful, to boom off and save the salmon fry hatcheries west and north of the wreck; and about the heartbreaking destruction that the oil had wrought.

And then Jerry and the rest started to asked *me* questions. That was not the idea at all; I was supposed to be asking the questions. But I had let the cat out of the bag with my remarks about Sullom Voe, and they wanted to know about it. I had crossed the wavy line between being a reporter and being an actor in the events I had come to report. After what I had heard, it seemed the only thing to do, whatever professional ethics might dictate.

So I told them. Over the next four days, as another Gulf gale blew in and kept me stranded in Cordova, we drew up a checklist of how they did it in Valdez before Good Friday 1989 and how the same companies had been doing it at Sullom Voe for the previous 10 years. Negligent tanker skippers, it turned out, were not the only hazard to which Valdez had been exposed all this time. We get them at Sullom Voe, no doubt, from time to time, but on a properly-run ship there should always be a qualified mate who has passed a pilotage examination for the port in question and who can take over and navigate the vessel on his own in safety. On *Exxon Valdez* that night this was not the case.

Our safety checklist went like this:

Item	Procedure	Valdez	Sullom Voe
1	Random aerial surveillance of all tankers entering and leaving port, up to 200 miles out:	No	Yes
2	"No-go" areas for tankers and a 10 mile "limit" around environmentally sensitive areas:	No	Yes
3	Compulsory reporting-in by tankers of course, speed, position and Estimated Time of Arrival when 200 miles from port:	not in such detail	Yes
4	Voyage instructions setting out the special requirements of the port are incorporated in charters for oil cargoes:	No	Yes
5	Commercial sanctions imposed by the seller of oil if the above instructions are disregarded:	Not normally	Yes

Item	Procedure	Valdez	Sullom Voe
6	Pilots board inbound tankers and leave outward bound tankers when beyond first reef in approaches to port:	No	Normally
7	Helicopter regularly used for pilot boarding and disembarking when sea too rough for pilot cutters:	Not normally	Yes
8	Full radar coverage of approaches from first reefs encountered:	No	Yes
9	Video tapes of radar plots are retained by port control for use in investigations after an incident:	No	Yes
10	Wind limits imposed on berthing:	Variable, about 50 knots	30 knots
11	Wind limits imposed on loading:	No	Yes
12	Visibility limits on berthing:	Varies	Yes
13	Four tugs for every ship berthing:	Maximum 3, usually 2	Yes
14	Voith-Schneider tractor tugs:	None	Two
15	All loading jetty mooring dolphins, as well as jetty heads, are fully fendered:	No	Yes
16	Computerised "rogues gallery" of offending and sub-standard ships, covering entire vessel casualty history, worldwide:	US only	Yes
17	All tankers boarded and inspected by qualified marine officers:	No (about 25 percent)	Yes
18	All port control officers in charge of tanker movements in port hold an appropriate skipper's ticket:	No (most are Coast Guard officers who do not require such tickets)	Yes

(Note: Alyeska no longer even required their marine supervisors to hold a ticket).

Item	Procedure	Valdez	Sullom Voe
19	Tankers with even minor defects (such as a missing rung on a Jacob's ladder) can be refused berthing until brought up to standard:	Unclear	Yes
20	All oilspills, however minor, and all breaches of safety regulations, are logged and made public each month by the port authority:	No	Yes
21	Marine pilots also carry out port control, radar monitoring and ship inspection duties:	No	Yes
22	Pilots are salaried employees of the port authority:	No	Yes
23	Federal/national government marine surveyor on call 24 hours, 25 minutes from terminal:	No (nearest may be in Seattle)	Yes
24	No ship allowed to berth unless oxygen level in empty cargo tanks eight percent or lower:	Not strict	Yes
25	Inventory of oil spill containment, recovery and clean-up equipment:	Poor	Good
26	Back-up assistance for spills over 2,000 tonnes available within 24 hours:	Doubtful	Yes
27	Permanent booms to protect environmentally sensitive areas and stop most oil escaping through the narrows:	No	Yes
28	Joint industry/local authority groups to monitor environment:	No	Yes
29	Specialist technical advisory groups, with elected public representatives present, to assess oilspill clean-up plans, harbour procedures and regulations on visiting tankers:	No	Yes

Item	Procedure	Valdez	Sullom Voe
30	All oil loaded via storage tanks after being stabilised:	No (direct loading of "live" or "spiked" crude is normal)	Yes
31	Sludge disposal	Shipped to "Lower 48"	Local firm

I do not wish to be unfair to those in charge of the port of Valdez, nor to exaggerate the preparedness of Sullom Voe for a disaster, but I have checked these points with a representative of the Alyeska Pipeline Service Company Inc., with the US Coastguard, with the Alaska Department of Environmental Conservation (DEC) and with the port authority at Sullom Voe. The checklist appears to be accurate. It is fair to say that precautions at Valdez since the spill have been exemplary; it could probably now cope with a medium-sized spill as well as or better than Sullom Voe, although neither could handle another big one satisfactorily without massive help from outside. Sullom Voe, like Valdez, is buying more radar cover and more containment and recovery equipment for oilspills.

The point, however, is that for ten years Sullom Voe was much better equipped to *prevent* such a disaster happening and had a range of deterrents which made everyone at the terminal hyper-conscious of safety. As I write this, Sullom Voe port control have just confirmed that a minor spill occurred a few days ago when a tanker loaded oil too quickly, resulting in a spill on deck. The scupper plugs had been checked before loading began, so only a few gallons were spilled into the sea, where they were quickly cleaned up. There are no grounds for legal action in this case but an unpleasant letter is on its way to the ship's owners in Geneva, asking them to make inquiries and to instruct their masters to take it easier next time. BP will agree to ban the ship if port control is not satisfied that she meets their exacting standards. In another recent example, a tanker master was fined for falsifying records to show that he had "inerted" his cargo tanks when in fact they contained more than the permitted oxygen level. Port Control at Sullom Voe frequently initiates successful prosecutions; at Valdez there have been very few. The principle, enshrined in the Sullom Voe regulations since 1979, is that safety and a clean environment must come before saving cash by making a quick getaway from the jetty.

Most tankers carry a reference book on the bridge, detailing the navigational arrangements and safety requirements of all the world's

Captain George Sutherland, the council port director who ensured that the lessons of *Exxon Valdez* were learned at Sullom Voe. Early in 1990 he proposed new radars to cover the blind spots on Shetland's rugged west coast and asked for more equipment to contain and clean up spilled oil in the port of Sullom Voe.
Photo: Jonathan Wills

Pilots on watch in the port control room at Sullom Voe, monitoring the movements of every tanker—from 200 miles out.
Photo: Malcolm Younger

major oil ports. It is published by an English company, Shipping Guides Ltd, of 75 Bell Street, Reigate, Surrey. It lists Sullom Voe, alphabetically, some pages before Valdez, and goes into great detail about what is expected there. On paper, the ports did not look all that different before Good Friday 1989. Where they differed was in the standard of professionalism and the severity with which the regulations were applied. By no means were all of the "good housekeeping" rules peculiar to Sullom Voe. As well as having its own special regime, drawn up in the oily wake of *Esso Bernicia* during that bitter January 11 years ago, Sullom Voe has been kept clean by rigid adherence to existing international regulations. They are not perfect yet, but it helps if you follow them. Not only was Valdez run with a pollution-prevention system inferior to that at Sullom Voe, but it also appears to have been less stringent in its application of the international marine regulations, a fault which has something to do with American reluctance, for financial reasons, to ratify all the international conventions on marine safety.

Some of the evidence comes from the Commander Woodle I mentioned earlier. After he had been in charge of the US Coast Guard at Valdez for some years, he accepted an offer from Alyeska to come across the harbour and join them as marine superintendent for the terminal. It was an unhappy but instructive career move. In 1984 he left Alyeska, to join another company in Seattle, after rows about the need, as he saw it, to clean up Alyeska's act at Valdez.

A month after *Exxon Valdez* stranded, he confirmed the details of the checklist I had drawn up with the Cordova District Fishermen United. And there was more. He told me: "BP officials I've spoken with generally held up your [*Sullom Voe*] terminal as a model and were appalled at conditions at Valdez". Not appalled enough to use their muscle in Alyeska to change things, it appears, nor to put up the cash to implement the changes.

More and better tugs were proposed for Valdez years ago, he said, but were turned down. Annual oil pollution drills were "a very canned performance"—a view echoed by the Alaska Department of Environmental Conservation. Alyeska always had prior warning of them and the drills were, as a result, "programmed". Much the same criticism could be made of Sullom Voe, to be fair, because a month before the November 1989 emergency drill at the Sullom Voe terminal even I knew its name—"Red Dawn". Real surprise drills are few and far between.

Commander Woodle could not remember a spill drill out in Prince William Sound, where the *Exxon Valdez* met her nemesis. "It was due to the expense and the risk", he said. As for the terminal's plan to contain a spill of 200,000 barrels, which was only slightly less than *Exxon Valdez* spilled, he said: "It was a total pipe dream. We never had the equipment or capability to do that".

Internal memos from the Alaska Department of Environmental Conservation, published by the *Anchorage Daily News* in late 1989, showed a catalogue of near disasters with tankers, failures of oilspill drills, and repeated warnings from the DEC's watchdogs—men like Dan Lawn, the boss of the DEC office in Valdez and the scourge of sloppy Alyeska management. What Dan Lawn and his colleagues had to say, when they went public in the *News* articles, bore out what Commander Woodle had told me a month after the *Exxon Valdez* disaster.

Several times, he said, it had been proposed to build a third radar station, giving better coverage of the 90 mile long Prince William Sound until tankers went past the Seal Rocks at the mouth of Hinchinbrook Entrance and out into the Gulf of Alaska. The extra radar was needed because of icebergs calving from the Columbia Glacier. *Exxon Valdez* was dodging ice when she altered course to switch lanes—but never altered course back into her intended lane. "It wouldn't have been expensive", Commander Woodle said. "But it was never installed". You don't have to have a skipper's ticket to understand that, if it had been installed, it would have seen *Exxon Valdez* heading for Bligh Reef; the Coast Guard would have had at least five and possibly 10 minutes to alert the ship by radio. If she had been a hundred feet to one side she would have missed the rock. Those are the figures Commander McCall gave me from his port captain's office in Valdez in May 1989. A 1988 incident in Sullom Voe, when a straying tanker was detected by harbour radar and her captain alerted to the danger of hitting Lamba island, makes the point.

But, instead of expanding radar coverage, the Coast Guard at Valdez had actually downgraded the radars they had, to save on the federal budget. It is not recorded that BP, with all their experience in Shetland, objected to this economy on the grounds that it would be a hazard to navigation. Nor did they (or the tankers' insurers, for that matter) complain when the original Valdez practice of piloting tankers all the way through Prince William Sound, rather than just to Rocky Point, was also discontinued.

Shetland's decision to go for fully-qualified port control officers from the start was known in Valdez, Commander Woodle confirmed. They also knew about a similar system in Canada. As early as 1981 he had asked the US Coast Guard for vessel traffic system staff to have supertanker masters' qualifications: "My proposal was that qualified pilots should take over port control. They know the ships and the conditions and can do a far more professional job than the people we had there. I've nothing against the Coast Guard. They do the best they can. But they do not have the hands-on experience."

If Commander Woodle knew all this, so did BP, Exxon and the three other Alyeska partners who also had shares in the Sullom Voe operation—Mobil, Unocal and Amerada Hess. But BP had fewer excuses than most. I do not know why they boasted about their "Rolls-Royce of ports" in Sullom Voe but failed to implement in Alaska the obvious, practical and sensible precautions to which they had agreed voluntarily in Shetland. Maybe their partners were too mean; maybe there was just a failure of communications in a rapidly-growing global corporation which was to become by 1989 the biggest energy company in the US. But I do know what a Cordova fisherman told me in the Club Bar, the night I met Jon Rush: "BP are treating America like it was a Third World country", he said.

Further evidence of these apparent double standards may be seen in the ballast water treatment and sludge disposal plants at Valdez. The Alaska Department of Environmental Conservation has repeatedly claimed that the failure of these two plants to do what they are supposed to do—remove waste oil and dispose of it cleanly—has led to chronic oil pollution of sediments in the Valdez Arm of Prince William Sound.

Sullom Voe also had teething trouble with its ballast water treatment plant but in the end it met the levels of oil in water (five parts per million) required by law. And at Sullom Voe the cleaned ballast water from tankers is discharged into a tide race which dilutes it rapidly; at Valdez it pours into an enclosed fjord with little tidal current. Sullom Voe's oily sludge is burned in an incinerator a mile away. Sludge from Valdez has to be shipped out to the "Lower 48" and the DEC maintains that lots of it has been getting into the Valdez Arm instead.

Then there are the problems at Valdez with air pollution, with the DEC again complaining about discharges of poisonous aromatic hydrocarbons to the atmosphere. On calm days the purple-brown haze from the terminal can be seen hanging over the fjord. Sullom Voe has had periodic problems with smoking flare stacks, but the terminal has a gas plant which reduces the aromatic hydrocarbon discharges to negligible amounts. Valdez has no gas plant; "spiked" crude is shipped out with the gas bubbles still in it, an infinitely more hazardous business than Sullom Voe's trade in "dead" crude—oil with the gas taken out.

The more we learn about the stewardship of Alyeska (and, by implication, of its controlling shareholder) the clearer it becomes that *Exxon Valdez* was just one of several disasters waiting to happen at Valdez and in Prince William Sound.

It wasn't all BP's fault; officials working in Valdez for the Alaska State Department of Environmental Conservation had been writing to their superiors, expressing worries about the terminal's preparedness to avert and clean up oilspills, as long ago as 1976, the year before it opened. Little was done about it. Why did the State of Alaska, the US Coast Guard and the Federal Government not have a regulatory system which was at least as good as what had been squeezed out of the oil majors by Consul Funkhauser's "tiny group of home-spun farmers" on the Shetland council? Surely they knew what went on at Sullom Voe. It had even been in the *National Geographic* magazine and the *Los Angeles Times*. The Alaskan and Washington politicians who gleefully accepted campaign contributions from the oil companies over the years, while agreeing to budget cuts for the Coast Guard, have some explaining to do.

The failure of state and federal regulation may partly explain, but does not excuse, what still looks like cynical inertia on the part of BP and their partners in Alaska. We may hear the details from the full audit of Alyeska, demanded by the Cordova fishermen in evidence to various Congressional inquiries and President Bush late in 1989. Meanwhile, Jerry McCune and I are working on the theory that it was a simple and rather nasty case of bucks before ducks.

So much for the environmental lessons which Shetland and Alaska can teach us, but what about all that money? We shall examine Shetland's financial tribulations in the next chapter, but first we may ask if they really did it better in Alaska, where income from oil to the state's Permanent Fund began to flow on 28th February 1977, a couple of years after Shetland's deals had been drafted.

What Alaska made out of oil was rather different from Shetland's experience, at least in scale. For a start, the North Slope oilfields were bigger, with production by 1989 at around two million barrels a day—compared with 1.2 million through Sullom Voe at the peak, (declining to well under a million by 1990). The onshore investment was bigger—the Alaska pipeline cost $11 billion, including the cost of the 12 pumping stations. That is about five times the price of the Sullom Voe terminal—but the huge investment offshore in the East Shetland Basin fields probably made the total cost of the Shetland operation even bigger than Alaska.

With a population of around 520,000 and a huge land area, Alaska dwarfs Shetland geographically. But there are some interesting financial similarities, even if the scale of things is different. In Alaska, as in Shetland, over four-fifths of the local administration's income was coming from oil by the mid-1980s; the various producing companies mostly had shares in the pipelines and learned early on to cultivate at least the appearance of friendly co-operation when dealing with local representatives who raised objections about the size of the community's rake-off and the employment of locals (as in Shetland, oil had quickly become the dominant employer); in both, the fall in oil prices during the mid-1980s became a useful argument when the oil companies needed to plead hard times and draw in their horns for a while. And in both communities there were frequent and outspoken rows about money and pollution.

We have already noted that the man who managed the construction of Sullom Voe between 1975 and 1978, Basil Butler, had been general manager for BP in Alaska, dealing with similar local organisations who wanted their slice of the financial action. Mr Butler went on to some of the top jobs in BP and was closely involved in many of the financial negotiations with Shetland Islands Council, particularly when things began to go wrong in the early 1980s.

Unlike the Shetlanders, the Alaskans gained a direct share of central/federal government income from the sale of offshore drilling licences, although British governments would argue that they have subsidised Shetland heavily in the past from such national revenues and continue to do so. But Alaska also has that Permanent Fund, similar in outline to the Shetland council's Reserve Fund and SIC Charitable Trust and, curiously enough, established in the same year, 1976.

A delegation of Alaskan politicians was among many from other world oil provinces who came to look at what Shetland was doing in the mid-1970s, and some of us who spoke to them then like to think that they learned from our experience. They certainly cornered more cash; by April 1989 the *Anchorage Daily News* could report, in among the oilspill stories, that the Permanent Fund had quietly passed the $10 billion mark. Six billion dollars of oil money had earned another six billion over 12 years, less two billion paid out in annual dividends to the citizens of the state from 1982 onwards. And the six billion dollars saved in the Permanent Fund represented only a fifth of the total revenues to the state—an amazing $30 billion between 1977 and 1989.

Not counting Alaska's other oil revenues, which meant that there was relatively lower local taxation than there would otherwise have been, the Permanent Fund's $10 billion assets in April 1989 amounted to a nest-egg of $19,320 for every man, woman and child in Alaska. In Shetland at the same time, the SIC Charitable Trust was valued at around £100 million. Converting this to dollars, at the autumn 1989 exchange rate of about $1.55:£1, we find that Shetland's trust was worth only $6,739 per head for Shetland's population of 23,000—not that the individual in Shetland or Alaska had access to anything like that amount of money. But that was what was in the kitty to earn income for the state and its citizens. Even when we take into account the sums in the SIC's Reserve Fund—a balance of £21 million by 1989, it is clear that Alaska amassed more than twice as much money per head as Shetland, from the same oil companies—and that is not counting some other revenues to the state. Someone in Shetland had blundered, even if we do take on board arguments about Alaskans paying more federal tax, because their incomes are higher, because the cost of living is higher.

Within eight years of the first oil income to the state in 1977, Alaska boasted a Permanent Fund of some $7 billion, averaging a return of 11 percent per annum and dwarfing Shetland financially as well as geographically. As in Lerwick, the Juneau, Alaska, cash nest egg must be kept intact, with only the income from the principal sum available for spending each year, after setting aside a portion of the interest to keep the fund growing in real terms.

In Shetland all the cash was put into funds controlled by the council, on the principle that the benefits should be spread as widely as possible. Charitable donations by the council have been made to many individuals who can prove need, such as the disabled and pensioners living on their own in substandard housing, but a large proportion of the annual disbursement goes to voluntary and public organisations of various kinds, rather than to individual citizens.

In Alaska, using a similar principle but different methods, part of the annual income from the fund is paid directly to the citizens—over $1,000 a head in some years, although down to some $800 (£470) by 1989. From a purely mercenary and personal point of view, that would have been very handy for me and mine. The annual Christmas bonus of £200 a head to pensioners in Shetland—later restricted to £200 for each household containing one or more pensioners or disabled people—was the nearest we came to this in Shetland. In recent years there has been talk of abolishing it, in favour of more selectively "targeted" charitable donations. In Shetland, not all pensioners are equally needy, and it seems likely that the bonus will, in time, be phased out.

The underlying philosophy of the Alaska deals was the same, however—to build up a fund while oil is here, to cushion the economic and social impact when it goes. The Alaskan fields will begin to tail off at much the same time—the mid 1990s—although, as in Shetland, the future depends on the market price of oil, back up to $30–$32 a barrel at the time of writing; on new inventions to extract it from the more difficult fields; on clever ways of extending the life of existing oil reservoirs; and, above all, on the government tax régime for the oil companies. Then there is the question of access to the oil which has yet to be developed; there appears to be plenty of it in both Alaska and the North Sea. Objections to opening up the undeveloped North Sea acreage are unlikely, whereas in Alaska, particularly after the Valdez spill, there is considerable opposition to drilling in the promising prospects of the Arctic National Wildlife Refuge, which the environmental author and former BP man Bryan Sage has described as "a pearl of inestimable value in the chain of refuges in the United States".

Seeing how it should be done, spy-in-the-sky pilot Chris Griffiths prepares for take-off at Sullom Voe with two visitors from Alaska. David Grimes and Rick Steiner (right) came to study Shetland's anti-pollution measures in June 1989, three months after the Exxon Valdez disaster revealed the shortcomings of Exxon, BP and partners in Prince William Sound, Alaska.
Photo: Malcolm Younger

Sella Ness—the port control centre for Sullom Voe and base for the tugs, pilot boats and pollution fighters.
Photo: British Petroleum Company

Like Shetlanders, the Alaskans are worried about the possible decline in population as the oil age wanes, although production from the Arctic fields is still rising. In both places the population rose by about a third in the first decade of the boom. Both have to keep these new people and their children if small communities are to retain a basic life-support system at decent standards. And that will cost big money.

I cannot speak for Alaska but it seems to me that Shetland has made a promising start in developing new industries and reviving old ones to take up the slack. Alaska has seen a lot of the state oil cash squandered in inappropriate and sometimes grandiose development projects, such as the huge grain silos which stand gaunt and empty on the shores of Valdez harbour, and what some saw as ill-advised agricultural schemes.

While Shetland could have learned a lot from Alaska about how to corner a larger slice of the oilcake, Alaska could still learn a little from Shetland about the benefits of municipal enterprise and the public support of small business. For example, rural schools and community centres in Shetland villages half the size of Cordova are much better housed and equipped, although Cordova is rightly proud of what it has.

The irony is that the most visible boost from oil to the economy of Cordova so far has been the Exxon money which flooded in to hire local fishing boats for oilspill clean-up in the spring and summer of 1989. And there were plenty of demands then for Exxon, the state and the federals to subsidise small businesses that faced ruin because their fishing boat customers were not buying nets, the tourists had cancelled their vacations, and the local employers' workers had all left to work for Veco, Exxon's principal contractor for the clean-up.

Whether they would be so keen on the subsidies which Shetland Islands Council makes to rural shopkeepers, I cannot say. Certainly, if they were available, there would be no question of keeping the details secret. Throughout the oilspill crisis, the daily briefings from Cordova City Hall and the Chamber of Commerce were models of frankness. The press sat in on everything, as did the oil company's PR liaison men, even when quite personal details of individual businesses were being discussed. That democratic, perhaps slightly naive, spirit of openness about public business may be the most important lesson that the Alaskans have to teach Shetland. And openness, or the lack of it, in public affairs will be a recurring theme in the remaining chapters of this book.

We have seen how the experiences of Alaska and Shetland contrasted and coincided. As a postscript to our digression along the 60th parallel, there is another coincidence. The man whom Exxon put in charge of dealing with the *Exxon Valdez* after her stranding, who supervised the offloading of her oil into other tankers, who had jury repairs effected in Prince William Sound, and eventually got her safely towed away for rebuilding, was their senior marine superintendent, one Captain Billy Duncan. He turned out to be a Scalloway man. His mother, who still lives there, proudly sent *The Shetland Times* a picture of him on the bridge. In the background I could see the mountains above Cordova. It is a small world, as Exxon, BP and their friends realised a long time before the rest of us.

The Money—Where it came from and where it went 7

Like the Norwegians, Shetlanders have sometimes been flattered by the media with the description "blue-eyed Arabs". The rest of Britain has been encouraged to believe that the islands are awash with petrodollars.

There is something in this. You cannot spend over £1,300 million on an oil and gas terminal without a little of the cash leaking into the local economy and, in a community of only 23,000 people, it does not take much of it to produce dramatic results. On top of the cost of the terminal, there was all the money spent on the new oilfield service bases, airports, houses, roads, schools, waterworks and hundreds of other oil-related contracts, large and small, over the past 17 years—plus the huge sums of money needed to maintain, operate and supply the Shetland oil economy.

So, even without the local council's "take" (as a result of its special Act of Parliament in 1974 and of its various deals with the oil majors), Shetland would have enjoyed a boom economy throughout the 1970s and 1980s. But the islands' special powers meant that the boom was enhanced and prolonged by the money paid from the industry to the people of Shetland through their local public authority.

Partly because of the complicated British system of local taxation, to which Shetland Islands Council is still subject like every other local authority, partly because of the ramifications of the British tax laws on charities, and partly because of the islands' various deals with a consortium of thirty North Sea oil companies, the finances of Shetland Islands Council are of Byzantine complexity. It sometimes appears that councillors themselves do not understand it all. But we shall try to make sense of it as best we can. The important result of all this complexity was that, by mid 1989, the council's nest egg of oil funds was valued at over £120 million, a sum which produces a sizeable annual return when invested, and which will continue to be boosted by annual payments from the industry until at least the year 2000.

The Shetland "oil money" comes from two main sources: "disturbance payments" agreed under the first deals with the oil industry in July 1974 outlined in the Disturbance Agreement; and annual profits from the tanker harbour which the council controls and operates as port authority, under the 1978 Port and Harbour Agreement and the 1974 Zetland County Council Act. That Act of Parliament contains a key clause which allows the council to set up a reserve fund. Without it, much of the oil income could have been "clawed back" in the form of reduced Government subsidies because it would have been treated like any other council income. With it, the council could embark on an ambitious programme of assistance to business, fisheries and agriculture, on a scale never before seen in Shetland (or anywhere else with such a small population).

In the first decade following the signing of the Sullom Voe Association Agreement (which we shall examine later) in April 1975, Shetland Islands Council received £62.4 million under the special agreements with the oil companies. But during the same period the council was paid a further £110.5 million in rates, the local property tax, by the oil industry. So, of the total payments of £174.7 million, almost two-thirds came from rates which any local authority in Scotland would have been entitled to levy on any large industry within its boundaries. And, from 1989 onwards, the council is also earning rents from the industry of about £10 million a year, backdated to 1978, following a 15 year dispute which ended up in the highest civil court in Scotland. There are also relatively small sums each year from the council's share in the Sullom Voe tug company, Shetland Towage, a joint venture with a commercial firm.

The Disturbance Agreement, effective throughout the lifetime of the terminal, allowed for a fee to be paid by the oil industry to the council "in respect of the import of crude oil into Shetland [by pipeline] and compensation for disturbance caused thereby. The fee was in three parts:

(i) a pipeline capacity fee of 15 pence per barrel, at June 1974 prices, payable until January 1985;

(ii) a fee of two pence per ton (of 2,240 lbs) of oil, at November 1978 prices, index-linked to inflation until 1990 and pegged at 1990 prices thereafter;

(iii) a minimum annual fee of £750,000, at November 1978 prices, payable until January 2000, after which it was to stop.

What this amounted to was a royalty on the oil. It was all paid into a charitable trust, set up in 1976 to avoid central government taxation or "clawback". By 1986, after eight years of production, BP estimated that the industry had paid the council something like £33.7 million under the Disturbance Agreement. And that capital sum had grown, thanks to dividends and interest earned, until in March that year the trust's market value was already £69 million. Disturbance payments peaked quite early, in 1984, and have never exceeded the minimum annual fees guaranteed by the deal. But the charitable trust was expected to continue to grow until index-linking stopped in 1990, after which, with limited spending, careful management and prudent investment, it should be possible to sustain it at a level somewhere over £100 million indefinitely. And that is a conservative estimate.

The Port and Harbour Agreement, signed in November 1978, provided for another local tax on the oil companies. Income from the council-owned port is used to top up a Reserve Fund. This fund's main purpose is to maintain the port of Sullom Voe, but it makes so much money that there has been a regular surplus each year, totalling £14.3 million over the first decade. The surplus is used for direct funding of other Shetland harbours that have nothing to do with oil, for various economic development projects, such as helping new local companies to get started, and also to top up the SIC Charitable Trust from time to time.

The rates and the rent from the council-owned land at Sullom Voe must, by law, go directly to the general fund of the council. It is used to finance the things which Parliament says all councils must do, as distinct from the things which Shetland with its unique financial arrangements decides that it would like to do.

The income from rates would have been even higher but for several factors: the industry built as much of its oil terminal as possible in the open air, without a roof—thus qualifying for exemption from rates on "outside plant"; a Conservative government, friendly to the oil companies, changed the law several times and reduced the terminal's liability for rates; and the companies themselves applied for and won every possible legal exemption and concession on rates.

It can be argued that the council should have been aware that rating income (British property taxes paid to every council in the land) was likely to fall over the years and should have held out for more from the non-rates deals back in the 1970s. But Sullom Voe has for over a decade been Shetland's largest single ratepayer; while its valuation for rates was reduced, council spending was not. So the terminal's owners ended up with bigger bills for every pound's worth of plant liable to pay the local property taxes and, the oilmen would say, this largely negated the advantages their accountants and tax lawyers had won for them.

Back in 1972, the fear was that the oilmen might not pay anything to Shetland other than the rates due on the terminal and the rent on the development land—at that time still in the hands of the land speculators. But the oilmen were not stupid; they knew the power of the media and were eager to minimise bad publicity. That was why BP, Shell and Conoco agreed very early on that Shetland should get a little extra cash. The money being talked about in 1974 seemed huge to the average Shetlander in the street but to the oil companies it was small change. By late 1985, after eight years of production, they had only paid out sums equal to three days' worth of oil production from the terminal (at January 1985 exchange rates and oil prices). By the end of 1988, it was calculated that the cost of the Shetland deals to the oil industry, after tax allowances, had risen to the value of 11 days' production.

With the terminal regularly shipping out cargoes worth between £20 million and £40 million in a single tanker, they could afford to appear generous. And they went to great lengths to make it appear just that, as in their 1986 public relations campaign which at times had the council reeling on the ropes.

By late 1973 it had become clear that they were willing to do business. Early in 1974 the council began to investigate setting up the charitable trust so that most of the oil money could be kept clear of the taxman and the Government. In the same year the first discussions took place with the oilmen about a lease for the land which the council intended to acquire at Sullom Voe. Their position was clear—they would offer about £1 million a year, at 1988 prices, for the "bare land" value of the site but there was no way they would pay a rent based on the "development value" of Sullom Voe, not least because they, not the council, were going to spend all the money to build an oil and gas terminal on that peat-covered hill far away in the parish of Delting.

They forgot, or did not take seriously, a feudal and perhaps unfair relic which was still the law of Scotland: if I rent you an acre of boggy land worth a rent of £100 a year and you, as tenant, then build a £1 million factory on it, your factory becomes my property in law and I can, in theory, charge you a rent of thousands of pounds a year. A key question in the long wrangle was whether or not the council had at any time given up this feudal right and, if it had, whether the rent should be £1 million a year or £100 million. In the end the council got 10 times what the oil industry had offered in rent but a tenth of what councillors had demanded.

On 10th April 1974, the last objections to the Zetland County Council Bill were settled and the Bill became an Act—the law of the land. From that day on it was clear to the oil companies that Shetland's powers were a fact of life. Playing games behind the scenes with land speculators, to undermine the council's position, was not a sensible proposition. More significantly, the industry was now committed to building a joint-user terminal for about thirty companies with shares in the oil. That would save some money through pooled construction costs, but greatly increase the running costs—because of the extra accounting and laboratory analysis required to apportion costs and income from the cocktail of oil and gas which would be piped into Sullom Voe from a dozen oilfields. The oil companies immediately began scheming how to make the terminal as cheap as possible and how to minimise the sums which had to be paid to the council. They came up with some clever stratagems, ranging from legal nitpicking to plain, old-fashioned delaying tactics.

Immediately following the passing of the Zetland County Council Act, however, everything seemed to be going well, as the council and the oil industry fixed things up. Within a few weeks, Shell came up with an overall development plan for the terminal, the first hard information the council had been given, followed in June by the first "formal estimate" of the likely throughput of oil and gas—a key statistic because the council was to be paid by the barrel. That became a very contentious prediction as the years rolled on.

On 17th May 1974, Shell UK had sent the council a letter, making them an offer which, it appeared, they could not refuse. It made forecasts about how much oil was likely to flow through the terminal, forecasts which were later hotly disputed, as we shall see. It also predicted, on the basis of those forecasts of throughput, how much money the council was likely to get its hands on through the proposed Disturbance Agreement. Both Shell's partners, BP, and the council were later to reflect on the wisdom of the following sentence in the Shell letter: "...in formulating the proposals...we have been very mindful of the views...of the [council]...particularly of the desire to avoid a burden falling on the Shetland ratepayers as a result of oil-related developments". An appendix to the letter had further reassuring news: "Rates will be payable...on terminal facilities in stages as commissioned...and will be calculated in accordance with statutory regulations". This was an early example of an important-sounding oil industry promise that was in fact no more than a re-statement of the law.

The Shell letter clinched the deal. In June there was euphoria in the council chamber when Mr Ian Clark, the chief executive, reporting on the secret negotiations with the oil industry, assured members that "residents in Shetland would not have to bear the cost of oil-related developments", as the minute of the meeting recorded. His confident forecast was based on Shell's promises. Unfortunately for him, many islanders, and even some councillors, took this to mean that when oil came they would not have to pay any rates at all. This is the theme of a persistent local story which is still in circulation—that Ian Clark actually told a public meeting that there would be no rates, thanks to the oil money. No-one can ever remember when or where this meeting was held, and I have not yet found anyone with a note or a tape of its proceedings. But I live in hope.

On 12th July 1974, the euphoria peaked, as the marriage of mutual convenience between the oil industry and the council was consummated in the signing of the Disturbance Agreement in Lerwick Town Hall. The full text of the deal was not made public until 15 years later when, in October 1989, *The Shetland Times* "discovered" a copy of it and published it, even as councillors were still agonising about whether to release it to the press.

The 1974 signing was the occasion for much municipal hospitality, reciprocated in full by the smiling oilmen. Not for many years would the industry and the council again be on such good terms.

Mr George Blance, the council leader, recommending the Disturbance Agreement to his fellow councillors, said his only worry was that the proposals were "so generous as to prompt central government to intervene". He need not have worried. The Government probably knew exactly how ungenerous the proposals were and, besides, it wanted Sullom Voe up and running as soon as possible, so that the oil revenue could start pouring into the UK Treasury. It did not know or care much about Shetland, as long as David and Goliath stories were kept out of the newspapers.

The Disturbance Agreement certainly looked good, particularly if you believed Shell's optimistic 1974 forecasts of oil throughput. In the council's first local structure plan, based on Shell figures, it was expected that the terminal would handle up to four million barrels of oil a day, with a third pipeline a distinct probability. But that was before the extent of offshore loading of tankers direct from the production platforms became clear. Sullom Voe's peak production proved to be only just over a third of what Shell had told the council in 1974. Shell also expected first oil production to take place from Sullom Voe by 1976; in fact, it was the end of 1978 before the first oil flowed, and that lost the council two years' income from the royalty clause.

In 1985 a council document declared that, because of Shell's inaccurate forecasts, plus higher inflation than expected, and the industry's efforts to seize every rating and fiscal concession available, the payments to the council were unlikely ever to rise above the minimum levels specified in the disturbance agreement. Worse, the council had lost the opportunity of reaping investment income from the higher payments which had originally been anticipated. The Disturbance Agreement, the council concluded after eleven years' bitter experience, had "fallen short of expectations" and needed to be re-negotiated. As Mr Malcolm Green, the council's director of finance, put it in late 1985: "In cash terms, we've got a third less than we thought we should get".

By 1985, no councillor agreed with the late Mr Blance that the Disturbance Agreement had been "generous". The oil industry's view was that you could not really expect accuracy from forecasts of throughput made four years before the terminal opened; that the companies had updated the forecasts as new information became available from offshore; that there were indeed provisions to review the deal in "altered circumstances"; and that everyone had agreed in 1974 that the deal was "fair and equitable". They maintained to the last, however, that the circumstances had not materially changed and in late 1988 the council was eventually forced to concede the point.

As we have noted, these payments were "compensation for disturbance" caused by the oil terminal. The expected disturbance would be to the unquantifiable "way of life", to traditional employment patterns, to local employers competing for labour for the first time in their lives, and above all to the council for the huge extra costs it would incur in providing houses, education, water, drains and social services to the thousands of new people who would settle in Shetland because of oil.

To put this into perspective, by the end of 1985 the disturbance payments had amounted in total to £27.3 million, the seed corn for the SIC Charitable Trust. Yet by March 1985, the council's general funds had run up a capital debt of £140.7 million, including debt on the four jetties which they built and controlled at Sullom Voe. That was a level of public debt per head of population more than three times the average for Scottish local authorities. So the agreement was compensation for disturbance but by no means for all of it.

The oil companies were on site by 1974, but at that time the council had still not bought the land from the speculators and, more importantly, it had not issued a lease to the companies. Early in 1975 a draft lease was presented to BP and Shell. Their reaction was to stall. But negotiations continued on setting up the administrative structure for the terminal and, on 21st April that year, these arrangements were embodied in the signing of the second major deal with the oilmen—the Sullom Voe Association Agreement (SVA).

This deal involved no cash for the council, apart from the rent of the land, and that was not to be settled for another 14 years. But it set up a non-profit-making company called The Sullom Voe Association Ltd, to "control, supervise and generally organise" four main areas of work:

- the design, construction, operation, management and maintenance of the terminal;
- the leasing of the land;
- making sure that the development met the requirements of British planning laws;
- concluding the remaining agreements between the council and the companies.

Crucially, the SVA Agreement obliged the council to make available enough land, at a "reasonable" rent and with provision for re-instating the site when the terminal eventually closed down. It also obliged the companies to honour the Disturbance Agreement, apparently a superfluous condition, as it was already a binding legal agreement; to pay for the whole cost of the terminal; and to conclude a deal with the council for a lease of the land.

A key clause was No. 17(a) which, the oil industry subsequently argued, meant that the council gave up those feudal rights in Scots law to a "developed land value" rent. This said: "For the avoidance of doubt, it is hereby confirmed by the parties that the ownership of and exclusive right to the terminal (excluding land but including fixtures not the property of [the council] or the harbour authority [*also the council*]) shall be with the pipeline groups [*the 30 oil company partners*] and not with the Sullom Voe Association or the council".

That seemed clear enough but each side had its own conflicting interpretation, a conflict that was to cost over £2 million in legal fees

Sullom Voe Oil Terminal–Organisational Relationship

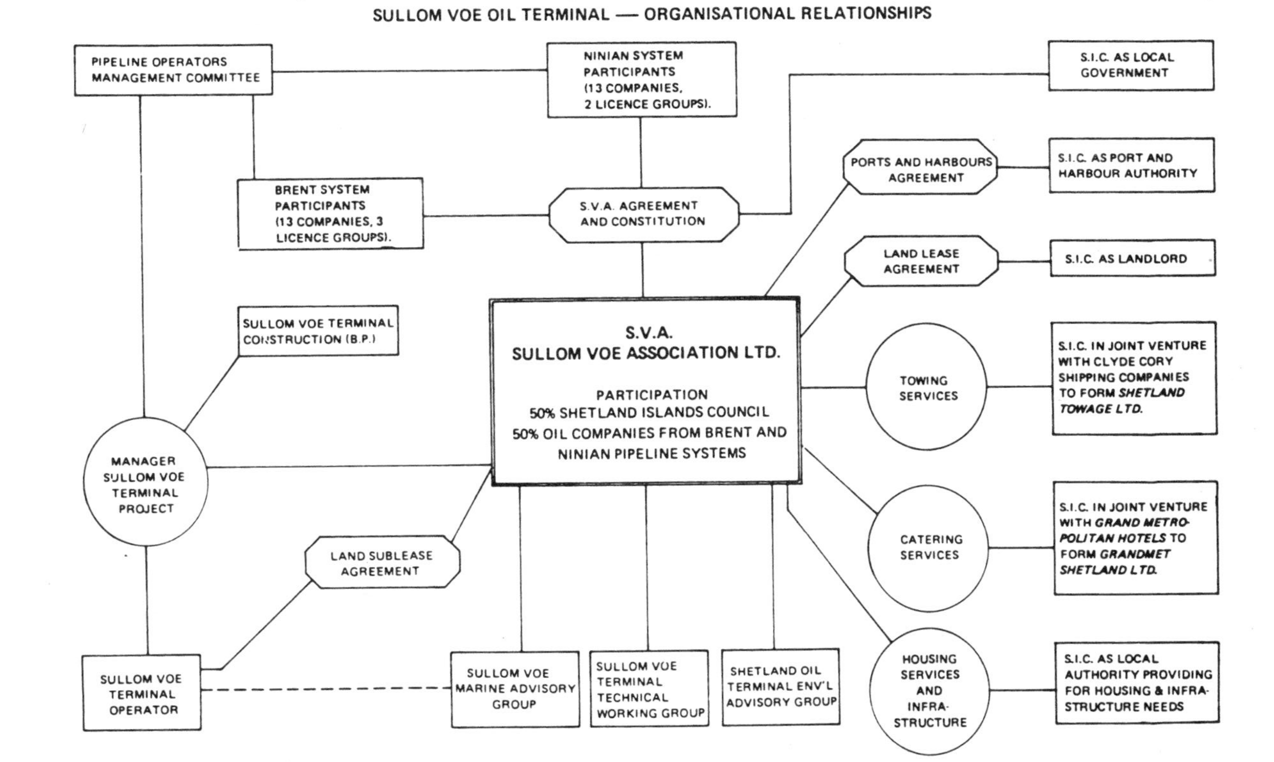

Credit: Reproduced from *Shetland's Oil Era* by Elizabeth Marshall with the permission of the publishers, The Research and Development Department of Shetland Islands Council.

and related expenses by 1988. Not for the last time, the lawyers for each side had failed to make themselves clear.

The signing ceremony was an optimistic affair. A top Shell executive, Mr Peter Baxendell, flew to Shetland from London and announced that the deal provided for "joint management" of the operations at Sullom Voe, with the council having "a controlling voice in all those affairs which affected the quality of local life".

The SVA Agreement, Mr Baxendell assured us in those heady days of 1974, provided "financial guarantees of performance" which went "beyond what was normal commercial practice". This was taken to mean the obligation on the oilmen to pay what they had already legally agreed to pay under the previous year's Disturbance Agreement. Big deal, was my somewhat cynical reaction at the time.

Mr Baxendell was waxing lyrical: "a unique partnership" had been forged between the council and the industry, he told the assembled dignitaries in Lerwick Town Hall. George Blance, less sanguine than he had been in the negotiations of a year before, was polite but commented that it would take "mutual trust" to make the SVA deal work. How right he was.

Mr Blance and his colleagues had to fight hard for even the limited control which the new agreement gave the council. There were last-minute, and still undisclosed, hitches, apparently only resolved by the intervention of the Labour energy minister, Mr Eric Varley. Mr Varley missed the signing ceremony, because of a combination of fogbound airports and an urgent House of Commons vote. But it was an important event for the Labour Government, then struggling for survival against a largely hostile House of Commons and desperately in need of oil taxes to ward off the International Monetary Fund. Mr Varley had a real interest in seeing that the Shetlanders were placated and it was believed locally that it was made clear to senior councillors that their special powers might be swept aside if they obstructed progress at Sullom Voe by nitpicking about clauses in this and other deals.

The great advantage to the oil industry was that the Sullom Voe Association took the heat off them in the crucial business of applying for planning permission, where many of the biggest rows could be expected. It was the SVA, not the companies, who put in the applications which had to be passed before work could legally begin. Council planners were told to co-operate as far as possible. The passage of the plans was to be smoothed, in return for what the industry had already agreed to.

The council, by falling for the offer of "a unique partnership", found itself considering detailed planning applications for Sullom

Voe from a body of which it comprised half the membership. The rows took place behind closed doors in the secret sessions of the SVA, which has never held a public meeting in 16 years, never published minutes of its proceedings and has only rarely, usually under local media pressure, even hinted at what its agendas contained. By the time detailed plans for the site layout, buildings and installations saw the light of day in the council chamber, the fighting was over, apart from sporadic attacks by a few maverick councillors who could not bring themselves to believe that inherent conflicts of interest really had been resolved.

This is not to say that the council was completely acquiescent. Council planners such as Mike Fenwick, the first planning director, later to follow his boss, Ian Clark, to Britoil, did their utmost to make sure the terminal was properly designed, safely constructed and that its visual effects on the landscape were minimised, but they had lost an important weapon in these battles—the exposure of the plans to public scrutiny at every stage of the process.

From the day the SVA Agreement was signed, the council became involved in a culture of secrecy and mutual dependence with the oil industry. It set the pattern for future years and contributed largely to the growing public scepticism about the council being bedfellows with BP and Shell. Common sense dictated that there must be conflicts between public and commercial interests. By becoming a "partner" in a secret organisation, the council forfeited some of the public's confidence in return for enjoying oil industry confidences which often proved illusory.

The council was not an equal partner. On paper, its two representatives and their expert advisers could not be out-voted on SVA by the two oilmen who represented the Brent and Ninian oilfield systems, through their twin pipeline groups. In practice, the company men, if they so chose, had a monopoly of the important and significant information, information which councillors might not have understood even if they had access to it. And, after the departure of Ian Clark in 1976, there appears to have been no-one on the council side who would strain the mutual politeness to the point where they said: "No, we don't believe you. Let's have an independent assessment, please".

The failing of the Sullom Voe Association was that there was no independent chairman and no provision for the council's technical and legal experts to see all the information. The oil industry successfully argued that some facts could never be shared because of commercial confidentiality. And, if the council representatives did see such information, they were bound by confidentiality not to reveal it. It was all a bit like being in the freemasons, which several of the key players on both sides undoubtedly were.

The wonder of it is that the council managed to use the SVA at all when it needed to make a stand. When it did so, as in the case of the new pollution control measures after the 1978 oil spill disaster with the *Esso Bernicia* tanker, it invariably had to go public, hazarding its special relationship with its oil company "partners".

On 23rd June 1975, the Sullom Voe Association was legally incorporated and the way was clear to complete the Sullom Voe terminal. Offshore, things were speeding up. Amazing and highly successful new-technology pipelaying barges completed the two main pipelines to Shetland by the summer of 1976—causing some to wonder if it might not have been cheaper to pipe the lot straight to Scotland at £1 million a mile and avoid all the hassle with the local council. Councils on the mainland would have fallen over backwards to welcome the oilmen.

The pipelines were in place but still there was nothing like an oil terminal ready to receive them. Back in Lerwick, council lawyers and finance experts worked on the fine print of the SIC Charitable Trust, trying to find a way in which its intended use "for the benefit of the inhabitants of the Shetland Islands" could be interpreted so that the cash could be used to fund economic development as well as purely charitable objects. The big problem was the power of investment. By law, trustees were obliged to invest in safe bets—government securities and "blue chip" companies quoted on the stock exchanges. But the Shetland plan was to invest some of it in local companies which were anything but safe bets. It became obvious to the experts that the charitable trust would never be able to perform as a vehicle for rural development, without putting some of the cash into other funds. They could not make the cheap loans for which local business was clamouring. If they took risks with these funds or gave money to non-charitable organisations they would be liable for tax, but at least they would be free of the encumbrances which Scots lawyers had hung, quite understandably, around the operations of purely charitable trusts. After much agonising, and with some of the key questions about it unresolved, the Shetland Islands Council Charitable Trust was legally constituted in September 1976. Three months later the UK tax authorities granted it full charitable status. At last there was a basket to receive Shetland's growing nest egg, which by then amounted to over £2 million.

Meanwhile, the man who had put the complex deals together, Ian R. Clark, had decided that his work in Shetland was done. He accepted a job in Glasgow with the British National Oil Corporation,

Labour's ill-starred attempt to set up a state oil company along the lines of Norway's Statoil; subsequently BNOC became Britoil, was "privatised" and later bought by BP. Just why Clark chose this time to leave has never been clear, but perhaps he decided it was best to quit on a high. Perhaps, also, he had become tired of the ignorant bickering which was the main contribution of some of the less intelligent councillors who gave him his orders. Certainly, by leaving when he did he missed some of the most bitter wrangles within the council and between it and the industry, wrangles which might have been better and earlier resolved, in my opinion, had he stayed. After his departure, the council never again showed the determination it had possessed in the early years of oil, and, whatever Clark's faults, it never again had a leading figure who had thought things through as clearly as he had or who had the moral courage to face down the oil industry.

There were those who envied and resented Clark, and, as soon as he had left, a whispering campaign began which surfaced periodically over the years whenever the going got tough. It was Clark, they said, who had fouled it up and then had run out and joined the oil industry. But this ignored the difference in philosophy and purpose between the state-owned BNOC and Shell/BP, which, in those idealistic days, was considerable. And perhaps, they whispered, that rumour that he had been on the take from Shell was not so far out after all. This was a reference to two London *Times* journalists, who arrived in Shetland to make it known that they had seen a copy of a cheque from Shell to Clark for some £25,000. In the summer of 1976, at the office of *The Shetland Times*, they promised me that they would produce the evidence. They never did. Those who knew Clark could not believe that he would take a bribe, but could well understand why the oil industry might have wished to buy off, or at least to discredit, such a formidable opponent. Knowing what we now know of the activities of MI5 against the Wilson Government in those years, one explanation for the London *Times* bribe story was that it fell into the same category—unauthorised attempts to destabilise those who challenged the established commercial order, which Clark appeared to do (rather more so than Harold Wilson, in fact). Or perhaps the attempt to smear Clark was aborted when some clever fellow in BP or Shell fell upon an even better idea. And was it entirely coincidental that Times Newspapers and Shell had a shared director?

There is an alternative explanation of the Ian R. Clark story, preferred by some journalists who have covered Shetland's oil age since the start; this holds that he was indeed very clever but not quite as clever as the oil industry made him out to be. According to this theory, it was no accident that the famous telex from Richard Funkhauser, the US consul at Edinburgh, was leaked to the press. An extract from this telex appears at the head of Chapter One of this book. I have been told, but I cannot prove it without revealing a source, that it was leaked by Funkhauser's own office. As we have seen, it portrayed Clark as the man who took on the industry and won.

According to my more cynical colleagues, it suited the oil companies to build Clark up as the wily "hornswoggler" who forced the oil industry into a costly deal, against their better judgment. Shetlanders were persuaded that they had won a famous victory whereas, as we have since come to suspect, the victory was hollow. Shetland could have won far greater benefits. Those in charge of the council at the time that the deals were done have always blamed pressure from central government as the reason why they did not hold out for more. In successive special pleadings to be treated favourably under UK laws on council funding, they argued that they had sacrificed some of their own local interests in the greater national interest of getting that oil ashore. Clark may have done wonders and it would be churlish to belittle his achievements, but Shetlanders did receive less oil cash per head than Alaskans did at about the same time. I suspect that the truth lies much nearer to the cynical interpretation than to the foolish theory that Clark was a sell-out. Certainly the oil industry had realised very quickly that few of the councillors who employed Clark were in the same intellectual league as their chief executive.

On 14th October 1976, a secret and significant meeting took place at Voe House, a former laird's mansion used by BP for wining and dining (and today the "official residence" of the terminal manager). Voe House lies at the head of a narrow and picturesque inlet on the Atlantic coast, a few miles from Sullom Voe. In mostly treeless Shetland it is famous for its trees. It is also famous for the "Voe House Letter", signed by the oil companies and the council at the conclusion of that "confidential" meeting. "Confidentiality", by the way, is the euphemism preferred by oil industry and council bureaucrats for secrecy. They also use the words "private", "in camera" and "closed" as synonyms, but on *The Shetland Times*, irritatingly, we always preferred to say "secret" when that was what was meant.

For 16 years the Voe House Letter has remained a secret. We know that it is highly significant. From time to time the oil industry has played down its significance, always a sign that a journalist is getting "warm". Extracts from it have somehow found their way into my files.

One of the clauses in a secret council briefing paper in 1985 read as follows: "*Other agreements and understandings*: Most significant among these are [sic] the Voe House Letter... in which it was agreed jointly between the oil companies and SIC that the terminal site lease would run concurrently with the Ports and Harbours Agreement [*then being drafted*] and that any extension beyond 31st August 2000 would entail adjustment to the fees payable under the Disturbance Agreement; and [*it would also run concurrently with*] the Occupation Licence... first issued by SIC on 23rd August 1978..."

No doubt the Voe House Letter said more than that but the significance of this clause, the council was to argue, was that it showed that the oilmen expected to have a lease and recognised that they would have to keep paying both rent and disturbance money if they wanted to renew the lease beyond the end of the century. In view of what we now know about the likely duration of the North Sea oil industry, the Voe House letter may have great significance when the next round of negotiations begin in 1995, the year when people will have to start talking about what happens at Sullom Voe after 2000. It is now clear that the dire predictions made by the oilmen during their long fight with the council over rent were somewhat wide of the mark. For example, the Ninian oilfield is estimated by its operators, Chevron Petroleum, to have reserves which will last at least until 2008; and BP have recently made it clear that they intend to be busy in the North Sea well into the second quarter of the 21st century. New, smaller fields are coming on stream all the time. If they can pipe oil to Sullom Voe, which was long ago paid for, then they will.

After the Voe House meeting, construction work speeded up on site at Sullom Voe. Throughout 1977 there was intense activity on the site. In March the council concluded the purchase of 1,012 acres of land at Calback Ness and announced that the cost would be recouped in rent from the oil companies. Meanwhile, work went on, in secret as usual, on the details of the next major agreement between the two sides. The outline of the Ports and Harbours Agreement had been sketched as early as 1974 but it was left until last because there was no port to agree about until 1978.

On 15th March 1978, the agreement was signed. It runs until 31st August 2000, with an option to renew it every five years for up to another 50 years, a wise provision, as things have turned out. The deal made detailed arrangements for the construction of the port, work which was in fact nearing completion by the time the document was signed. There was provision for maintaining and adapting the jetties and other plant, as the needs of the terminal might dictate; a system to control pollution and to respond to emergencies was set up; and arrangements were made for financing the four jetties, which were to be owned by the council but paid for (maintenance included) by the industry.

Then there was the money. Lots more money, in three separate but related parcels: until the year 2000 the oil companies would repay to the council each year 2 percent of the estimated cost of building the jetties; the jetties would be kept in good working order at the companies' expense; and thirdly there was another "royalty" —a penny a ton, as distinct from the metric tonne, on all the oil and gas shipped over the jetties until August 2000. The penny was a 1975 penny, index-linked to inflation over the years.

This sounded good, but some of us could remember talk of a penny a barrel when they went into the negotiations with the oilmen. Somehow it had changed by the time that they emerged from those smoke-filled rooms. It would be uncharitable to suggest that some councillors did not know that there were seven-and-a-bit barrels of oil in a ton, and that they might have extracted seven times as much money as they did under this agreement. But there it was, signed and sealed. By 1985 the agreement had earned the council £22.7 million, to say nothing of a total surplus on the harbour operating account of some £14.3 million.

The money was in the bag. The only problem was—which bag? Because they were worried that the Government might try to get its hands on some of Shetland's oil cash, council accountants in April 1978 put large sums of Disturbance Agreement money into the Reserve Fund which they were authorised to set up, under the terms of the 1974 ZCC Act. It later turned out that they had been doing this since the first payments, in the financial year 1974/75. This, as the auditors eventually pointed out, when they unravelled the tangled web of the council's finances, was a potentially disastrous mistake. They should have put it into the SIC Charitable Trust.

Had they understood the small print of their own Act of Parliament, the councillors would have realised that the Reserve Fund could not be established until the harbour account at Sullom Voe was in surplus. That did not happen until the financial year 1979/80. As a result, several million pounds were sitting in a fund which did not legally exist. The Government would have been quite within its rights to say that in fact the cash was, or ought to be, in the council's general fund, and that the council therefore required that much less of a central government subsidy for the years in question. Worse, the tax authorities might decide that, because there was no charitable trust to receive the money until December 1976, anything in the "Reserve Fund" before that date was liable to hefty taxes.

The foul-up created quite a furore, with the local newspaper and radio station reporting bewildering new revelations of the details of the affair almost every week—and being castigated by councillors who thought the media should behave more responsibly and hush up such embarrassments in the interest of "Shetland"—their favourite line, then and now, when they are caught out.

After several years, and innumerable sessions with the Inland Revenue in London and the Commissioner for Local Authority Accounts in Edinburgh, the national authorities were persuaded to agree that the funds had always been held "in trust" for the people of Shetland and therefore were not liable for tax after all. Not a few questions were asked, such as why the council's eminent legal and financial advisers had apparently not understood this unfortunate problem, in return for the large fees which they were paid.

The Reserve Fund affair was a salutary warning of the dangers in a council trying to build up its nest egg, free from outside interference but within a framework of national laws. There were others: councillors, sympathetic to pleas for financial help for the local fishing industry, made what turned out to be an imprudent loan to a fishing boat crew from the charitable trust, of which the council is the sole trustee. The Inland Revenue men did not like this at all and, in studying their criticisms, councillors discovered that perhaps the charitable trust was not quite the answer to their prayers which they had imagined. The trouble was that the trust was not supposed to hazard its money by grants to financially shaky outfits like fishing boats. The capital was supposed to be kept safe and the income used purely for charitable purposes. Most of it was used for precisely that, with admirable results. But even the shakiest Shetland fishing boat partnership could not be called a charity, however much you stretched the English language. What was at stake was the trust's charitable status: just one uncharitable investment, grant or loan could leave the whole nest egg wide open for tax, as councillors were horrified to learn. Once again, the auditors and the taxmen were persuaded to accept that this had been a mistake. The charitable trust survived. The immediate consequence was that most grants to commercial organisations were subsequently made from the Reserve Fund, which was kept quite separate from the SIC Charitable Trust.

This left Shetland without the large, revolving loan fund which had been a cornerstone of their economic development policies. There has been talk of setting up a Shetland Enterprise Trust which would pay tax and thus be untrammelled by charitable trust laws. The charitable trust has now appointed new trustees in a complicated manoeuvre to get round Government controls on how it spends its cash.

There are strong arguments for easing up spending and leaving most of the money intact for when the big downturn in oil arrives. By 1989 the council had a serious problem in finding labour to carry out the various admirable projects it had in mind—particularly in the field of housing maintenance and repairs. Unemployment was less that 4 percent, half the national average. Contractors' prices were rocketing; they had the council just where they wanted them, and, as I write this, the council is faced with the problem of not being able to spend the current year's allocation of cash before the end of the accounting year.

Council accountants have had to be quick on their feet in these financial quicksands, and, however much the SIC advertised for accountants and lured them with glossy videos of island life, there were somehow never enough of them to get through all the work that was demanded of them. Some of the work had to be farmed out to commercial accountancy firms.

One of the novelties the council finance experts came up with was an entirely new concept—the "repayable grant". Not a loan, you understand, in case that broke some regulation or other; and not a once-and-for-all grant that could not be recovered and might breach some other statute; but a repayable grant. A loan, in other words. Thus far, the auditors have not complained and, if they don't, why should we?

The council has been only moderately successful in getting hold of the money. Such feats of acrobatic accounting raised doubts about its skills in keeping hold of it and applying it to the stated aims of sustaining the Shetland economy as and when the oil industry dwindled. Fortunately, perhaps, it appears to be a long time a-dwindling.

So Where Was the Graft? 8

So where was the graft and how big was it? The answer is that there was less than you might expect, and it was mostly on a very small scale. Are Shetlanders more honest than most folk, or are they easier to bribe? The truth may lie somewhere between the two.

It is difficult to convey the scale of the money that flooded into Shetland in the 15 years after oil arrived; there was the huge capital cost of the terminal and the Sullom Voe harbour; then there were the hundreds of permanent new jobs, at unheard-of wages; plus the spin-off effect on almost all local businesses, so that by the autumn of 1989 Shetland's unemployment rate was less than 4 percent, well below the British average and less than half the Scottish figure; then there were the temporary jobs when, in 1987–89, the oilmen had all that extra cash mending the bits of the terminal that had rusted.

On top of that, the demands of the offshore oil industry in the East Shetland Basin created steady jobs at Sumburgh and Unst airports and in the port of Lerwick, which sustained hundreds more people. Little wonder that the population of the islands, around 17,500 in 1971, had increased by more than a third—to 23,000—by the mid 1980s.

That population was mostly enjoying wages and living standards well above the British average, although pockets of poverty persisted (ameliorated to some extent by the miniature welfare state which Shetland Islands Council was able to sustain with its oil funds). Apart from a two-year dip after the 1985/86 oil price crisis, when the oil industry claimed that it was seriously thinking of leaving town, the fifteen year Shetland boom has showed no sign of slackening.

Oil is now easing up. The first boom is over and throughput at Sullom Voe is slowly declining, but it is clear that, accidents permit-

Pilot boats and tugs at Sullom Voe have meant 12 years of steady work near home for dozens of Shetland seafarers. In earlier times they would have been scattered throughout the oceans of the world.
Photo: Jonathan Wills

ting, the terminal will be operating well into the 21st century. And the service bases for the offshore industry have been enjoying a second boom, as interest quickens in exploration around Shetland and the new generation of small fields come on stream and, mostly, plug into the two Sullom Voe pipelines.

Meanwhile, the old mainstay—the fishing industry—is in crisis, not because of the oil companies, whose platforms, pipelines and seabed debris paradoxically create the only safe refuges for the fish to breed in peace, but because of gross over-catching of white fish by fleets from all over the European Economic Community.

The herring and mackerel boats now have a better future, thanks to severe and belated conservation measures, and this "pelagic" fishery is reviving. But other work has come along to take up the slack—notably the salmon farming boom.

By 1988 fish farming was worth more to Shetland than the entire whitefish or "demersal" catch, employing several hundred people, directly and indirectly. Some of the investment came from council loans but most of it poured in from finance houses in London and Norway who knew a good thing when they saw it—Shetland-farmed fish commanded a premium because of its quality, thanks to the cleanest seas in Europe, relative freedom from fish disease, some clever image marketing, and, not least, to the unique way in which the development of the new industry was controlled by Shetland Islands Council.

Ian Clark was a man of foresight, but not even he, surely, envisaged the ZCC Act of 1974 being used to control and sponsor fish farming. The clauses in the Act which allowed the council to license all works within three miles of the shore were used, 10 years after the law was passed, to regulate who could have a fish farm and where. Locals, or at least local front-companies, were given preference. Councillors claim that this has prevented overcrowding and the proliferation of get-rich-quick schemes which would enrich their shareholders and impoverish the environment of the sheltered voes and firths of Shetland. Conservationists are not so sure. The Nature Conservancy Council, a government body, has even accused the Shetland council of being less environmentally aware than its despised counterparts on the rest of the Scottish Coast, the unelected and, some say, autocratic Crown Estate Commissioners.

In Scottish law, the Crown owns the foreshore and the seabed; in Shetland, according to the salmon farmers, the foreshore belongs to the coastal landowners, under the ancient Norwegian Udal laws which once applied to Shetland and, they say, still ought to; they also claim that in Udal law the seabed out to the territorial limits, now 12 miles, is owned by no-one, although various parties, including the Government, have rights and responsibilities over it. The point has recently been tested in the courts. The Crown won but the outcome of an appeal by the Udal side is awaited with interest. What matters for salmon farming is that control is clearly invested in the islands council, by law, even if the Crown Estate claims rent for its use.

Not only salmon farming has boomed; the tourist industry, for decades woefully underdeveloped, has recently taken off in a big way—thanks to better air and ferry services, indirectly the result of oil, and to the sudden availability of bedspaces which used to be occupied by migrant oil-related workers from the south. Many of the visitors come to see the birds, the seals and the otters. The potential for expansion is enormous, if the European fishing industry, which has been destroying the small fish on which Shetland's marine wildlife depends, can be controlled in time.

Increasing numbers of professional people are finding that they can live and work in Shetland for at least part of the year, thanks to greatly improved telecommunications—recently improved to international standards for fax, electronic mail and digital transmission, again thanks in part to the demands from the oil-related industries. They appreciate the lack of traffic jams, crime and general hassle—Shetland, even busy old Lerwick, is still a refreshing contrast for refugees from the congested districts in the south. And the social, educational and medical services, provided by Shetland's municipal mini-state, make island living even more attractive. This new wave of incoming workers, mostly self-employed in the more cerebral professions, will make an increasing contribution to island life and finance. Only the appalling winters will deter large-scale settlement.

So prosperity seems to be here to stay, at last. And most of the money for the centrally-heated new houses, the new cars, the hi-tech leisure pursuits—Shetland's first jet ski was observed and noted by the Lerwick Pierhead Skippers in the summer of 1989—has been come by honestly, by hard graft. Graft of a different kind may exist but it is hard to spot.

One of the first questions was who should get the jobs at Sullom Voe. The answer was simple—it was mostly the Scots, Irish and Geordie "bears". It was a different matter when it came to the "jobs for life" on the pilot boats and in the Sullom Voe tugs. It may be a statistical coincidence, but a sizeable minority (a majority, by some accounts) of the jobs went to members of the three local masonic lodges in Lerwick, Brae and Unst. There may be a fourth lodge, in the island of Yell, but until I join I cannot be more specific and after this book I do not expect they will ask me to.

Many years ago, a lawyer who arrived in Lerwick from Edinburgh was asked what he thought of the community, after a residence of some six months. Well, he confided, when he had arrived he had been under the impression that freemasonry was a conspiracy. Now he knew better. For how could the majority be a conspiracy? Freemasons are, in general, not a majority of the male population of Shetland, although in some districts and islands they may well be, but they are extraordinarily numerous.

As in other places, the vast majority are entirely innocuous masons, joining because the declared beliefs of the craft appeal to them on a philosophical level and/or because they are attracted by the camaraderie of the excellent cheap bars and social facilities which the lodges offer. And they may have noticed that preferment in business or employment sometimes—and quite improperly, according to masonic rules—seems to come the way of otherwise unremarkable men who happen to be members of the lodge. But there has been no published instance of masonic graft and corruption in Shetland.

Masons in the islands have a long history. Lodge Morton in Lerwick was founded in the 18th century, one of the oldest in Scotland. It was a natural focus for the landless merchants, commission-men and public officials who took on the dictatorship of the old landlords and beat them at their own commercial game, although some 18th century lairds were also masons and enjoyed the card parties and the all-night drinking alongside their upstart brethren. By the late 19th century, Shetland freemasonry was powerful, respected and proudly displayed: the new Lerwick Town Hall was opened with full masonic honours in 1882, complete with a splendid stained glass window donated by Lodge Morton. Most of the Lerwick "establishment"—the merchants, shopkeepers, property owners and other small town bigwigs, were either masons or on good terms with freemasonry. It was the same in many other small Scottish towns.

The Shetland brotherhood was by no means confined to the grand and the monied—many poor men were members, not least because it might help them to get decent jobs when work of any kind was hard to come by. And, in an age when so many young Shetland men went to sea for a living, membership of a lodge conferred immediate advantages when they went ashore, friendless and poor, looking for recreation or another berth in foreign ports. Freemasonry was the norm, for the comradeship and practical assistance it offered to a young merchant seaman making his way around the world. It proved of benefit to the hundreds of Shetlanders who joined the armed services before, during and after the two world wars of the 20th century. Whether you were with the British Army of the Rhine or serving with the Hong Kong garrison, the lodge meant security, companionship and the friendly help of like-minded contemporaries. It also kept you out of trouble, helped you to meet nice, clean girls and put up bail when you fell foul of the law.

For the oil industry, one of the immediate advantages of BP taking over from Shell as the front-men in the early days of the Sullom Voe terminal, was that BP was, and is, one of the most masonic companies in Britain. BP employees were instrumental in setting up a thriving lodge, at the village of Brae, in the late 1970s.

If you're a local and you want to get on in BP, being a mason helps. Masons would argue that, since there are so many of them in Shetland, it is entirely natural and statistically probable that there should be so many of them at Sullom Voe and in many other workplaces.

Few in Shetland regard this as being at all odd or irregular. In the council itself, being a mason is no disadvantage if you want a job at their side of Sullom Voe. There are advantages in this: it means that the men are more likely to get on together in the close confinement of the pilot cutters and the long, boring night shifts on the tugs; they are more likely to know each other well, a circumstance which should help them to perform as an effective team when emergencies occur; and they are much less likely to go on strike or cause trouble.

British freemasonry has been severely criticised in recent years, with some allegations of graft. But in Shetland there are many other more or less exclusive societies. There are boating clubs, church groups, political parties, trades unions, angling clubs, Rotarians and Up Helly A' squads for the annual fire festivals in Lerwick and elsewhere; there are sports clubs of all descriptions and, at the last count, something over 700 voluntary organisations of various kinds. Shetland is a very clubbable place and, to confuse matters, membership of all these outfits overlaps, not least with the freemasons. There is probably even more freemasonry in Orkney. When I first went there in 1973, to become prospective parliamentary candidate for the Labour Party, I was shocked, in my puritan innocence, to discover that a number of well-known freemasons were respected members of the party and the trade unions in Kirkwall. The unusual thing, it turned out, was that there were so few in the Shetland Labour Party.

It is natural to favour those who share one's own opinions and experiences, so we should not be surprised that many appointments have been made during Shetland's oily years, and probably just as often before, where a candidate's qualifications for the job were not the only attributes considered. No, I do not like it. I'm not a mason and hold membership cards for only the Royal Society for the Protection of Birds, the Lerwick and Bressay boating clubs and the National Union of Journalists (now there is a conspiratorial minority, according to some Shetland councillors!)

During the hectic years of construction at Sullom Voe, all sorts of rackets went on. And you didn't have to be a mason to be in on them. It was easy, if you knew how, to smuggle out of the site large quantities of scaffolding planks, welding rods, nuts and bolts, bedding, electrical wire and fittings and even whole welding sets and bathroom suites. All this was immediately put to good use locally, of course, and probably to better use than if it had stayed to rot on the site—for the contractors were notoriously slapdash and tended to over-order, lest BP should find them short of essential kit, come the next big push.

The wastage, particularly in the construction camp canteens, appalled islanders who had been brought up to scrimp and save. Few felt any qualms about "liberating" food and materials which otherwise would have gone to waste, and many an illicit midnight feast was enjoyed by the relations of those who worked there.

BP and the major contractors came to be regarded as fair game. And few were caught at it—although I can think of one case of a man who was stopped by the security men on his way out of the gate, and found to have some reams of company notepaper in his car. He was not welcomed back by the management, nor by his mates. The companies seemed to take a far more serious view of drunkenness among the "bears" than of theft. Many a man was shipped back to Glasgow ("withdrawal of accommodation" was the euphemism used, and as there was nowhere else to sleep it meant the sack) for drunken incontinence, sometimes followed by a minor court appearance. They even kept a "wet beds register" in the camps at Firth and Toft.

The day that the Queen came to perform the official opening ceremony of the terminal in May 1981, the Provisional Irish Republican Army, who had a few sympathisers among the construction workers, planted a bomb in the terminal's power station. It was timed to explode as the royal party passed the building. It went off on time but the Queen was late and left without knowing anything about it. No-one was hurt and only minor internal damage was done to the power station. Some days later a local man who had worked for a time at Sullom Voe asked me about the bomb and, as I had been even later than the Queen, I was one of the few reporters who had heard it go off while she was making her big speech.

"And what size of a thing was it?" he asked.

"Oh, about seven pounds. The police are trying to work out how they smuggled it in through the security gate", I replied.

"Seven pounds? Man, that's nothing! You should see the weight of stuff we smuggled out of the place!" my friend replied.

All this did strange things to the Shetland psyche. No-one in the country districts had ever locked their doors in the old days. Few do now, except to keep out wandering, friendly drunks, if the family has decided on an early night. Not many folk locked their doors in Lerwick either, although lots do today and some have even taken to locking their cars. There was petty theft, but the culprits were usually detected. It had to be petty because there was nowhere to hide the loot if you took anything bulky. And the theft rate is still very low in the islands. Most of it is committed under the influence of drink and the detection rate is correspondingly high.

The mutual respect for other citizens' property is one of the attractive features of Shetland life and is quite remarkable when you consider how the current prosperity has greatly increased the number of valuables to steal. Yet BP and other big company property is now seen by some as a different matter. The same can apply to the property of the local government, which has more of it than ever before. A double standard is subconsciously set, often by people who would be shocked to be thought other than upright, law-abiding citizens.

There was also a little bit of graft in the way contracts were operated at Sullom Voe during the construction boom. It was not unknown for a van driver from Lerwick to make several unscheduled stops at private houses on his way to deliver consumables at Sullom Voe. This has probably declined since the operational phase of the terminal began.

Some will blame all this small-time crookery on the general decline in the public morality of the acquisitive society. Others blame the decline in influence of the churches, of which Shetland still has a great many, most with their Sunday morning traffic jam of expensive motor cars parked outside (cars, rather than drink and tobacco, are the sole indulgence of the religious in Shetland).

I prefer to think that BP and the other big organisations which came to Shetland brought this upon themselves. By their conspicuous waste they created the idea that they did not care, that their property was up for grabs—be it a rare case of embezzlement or the much more common theft of office pens and paper.

BP are not generally regarded in the same way as other individuals and organisations. After 15 years they have not really become integrated, much as people may like and respect individual BP employees who come to work in Shetland, and much as hundreds of local people depend directly or indirectly on the company for their living.

The fact that relatively few Shetlanders, even the masons, have made it to senior or even middle grade management at Sullom Voe is noted and quietly resented, even if the company is assiduous at hiring local apprentices and office trainees. Shetlanders have not forgotten that the "travelling men" who built Sullom Voe appeared to have much better leave and conditions than the locals who worked alongside them; islanders did not like the persistent "economy with the truth" which characterised BP's public relations statements over the years on all sorts of issues ranging from the first oilspill through various industrial disputes and petty restrictions on employees (such as the introduction of charges for formerly-free buses to work, which caused outrage) to fully fledged rows with the council about money and the environment.

Shetlanders may not have a great deal of respect for their council—most will have a tale to tell of its inefficiency, stupidity, timidity or greed—but it is still their council. BP is not their oil company, however much it may try to play Santa Claus and ingratiate itself with all sorts of charitable grants and free trips for the needy. The public anger over BP's persistent refusal to pay rent or water charges for the terminal for 14 years was not a figment of a local newspaper editor's imagination. The company's attitude caused deep offence, based on an indignant recollection that BP had been happy enough to throw money at Sullom Voe to get it built, but appeared to be extremely mean when it came to honouring their deals with the council.

Even if the councillors and their officials had not been up to it when it came to dotting the i's and crossing the t's of the agreements, the feeling remained that BP and the other partners in Sullom Voe could and should have been more generous. We got a lousy deal, the theory goes; we never did trust that Clark fellow and besides, the councillors could not run a candy stall; so what has been going on?

Alas for the conspiracy theorists, nothing so machiavellian was necessary for Shetland Islands Council to emerge with less cash per head of population than the Alaskans or the sheikhs. We are not talking about financial corruption, in the sense of "Here's some cash; now, this is what you have to do for it ... " In the years when they were building Sullom Voe, it was well known that there were often spare seats on flights to and from the British mainland. Because of Britain's strange system of preventing real competition between private airlines, the price of a ticket on a scheduled flight was astronomical; the temptation was to "fly BP", Shell or whatever. As a relatively impoverished young reporter in the late 1970s, I did my share of hitch-hiking out of Sumburgh and Aberdeen airports. Security was extremely lax—until the Dan Air disaster of July 1979, when 17 people died in a plane which ran off the end of the Sumburgh runway into the sea. Some of those on board were reportedly not on the passenger list, leading to questions from the insurance companies. Even after security was tightened, it was still not difficult to get on a charter plane if you knew the right people. It was harder to walk aboard a helicopter, even before the appalling Chinook crash off Sumburgh in November 1986, which killed another 45.

It was one thing for ordinary folk to take a free airplane ride, but when it was elected councillors that was another matter. Some councillors were outraged when, in the summer of 1989, *The Shetland Times* questioned them about the possibility of their having taken BP flights on private business and pleasure; it had never occurred to them that accepting such a favour from BP might compromise their dealings with a company which they were suing for hundreds of millions of pounds, or that it might at the very least give an unfavourable impression to the electors. Most said honestly that they had never flown BP and only one openly admitted doing it. It turned out that he had been on council business anyway—in which case, why was the council not paying? Councillors were embarrassed by the local press into holding an inquiry, secret, of course, into a free BP trip to London taken by their chief executive officer, Michael Gerrard. He was reprimanded and given a written warning.

From the start, BP was a generous host on the Shetland social scene. "Corporate hospitality" is the euphemism and no doubt at the dinners and dances much frank talking about council/oil industry business was done, and some good may have come of it. Councillors who might be fulminating against BP at a Town Hall session one day could sometimes be seen drinking BP's wine in the same building later in the week. It never entered their heads that they were doing anything out of the ordinary. In Shetland there is a long tradition of social gatherings being used to defuse personal animosities which could become dangerous to the community if allowed to fester. BP were just going along with a useful and enjoyable local custom. But the oil industry was not the crofter next door with whom you were feuding about a piece of boggy ground. This was the big time and many electors thought that councillors should have kept their distance—at least between themselves and the senior managers and negotiators on the oil companies' side.

The pathetic and rather endearing truth is that Shetland's councillors did not need to be bribed. Show them a grey dinner in a grey hotel, and a lot of grey men in grey suits with fancy accents talking to them as if they were equals, and they were captivated. The oilmen turned out to be human, just like them, and charming fellows when you met them informally. All very cosy, all very nice, and all very dangerous, as BP knew full well.

What we are talking about is a corruption of the intellect, a collective failure of self-confidence. It has its roots in a sense of being small players in a big story, fooled into thinking they had outwitted clever financiers and industrialists. It is fun when the TV documentaries portray you as David confronting Goliath, except when Goliath wins. Or maybe it is Cinderella with the Wicked Stepmother as heroine, Prince Charming a charlatan, and Cinders sent back to the scullery because she is too small for her glass slippers. Whatever the plot, the results are curious, to say the least, and a cautionary tale for the inhabitants of the world's other oil provinces.

Home Truths on Home Rule 9

So—we have arrived at our "place in the sun". Many must envy us, consider our problems minor and our continuing complaints as mere whining. After all, thanks to oil, Shetland is now one of the most prosperous parts of the United Kingdom. Despite its failings, the council has used its oil money to promote social and economic projects of undoubted benefit to local people. Surely everything in the garden is lovely? And we also enjoy more self-government than any other community except the Channel Islands and the Isle of Man.

"Enjoy" may not be quite the right word, for there is widespread dissatisfaction with Shetland Islands Council, which has shown itself to be obsessively secretive, sometimes unresponsive to public opinion, occasionally incompetent on a massive scale and often disliked and mistrusted simply because its activities now permeate every corner of island life.

Some of the essential preconditions for effective self-government are missing. For example, the electoral system bars thousands of citizens from holding public office. We cannot blame the oil industry for this, even though it came about at the very time that they arrived in Shetland. It is a result of the Conservative Party's "reforms" of Scottish local government in 1973, which the incoming Labour Government did not alter in 1974. Before that, employees of Lerwick Town Council were eligible for election to the old Zetland County Council, and vice versa. But, since they were merged in 1975, no council employee can stand for election.

Except in Shetland, Orkney and the Western Isles, which have "single-tier" councils, local government functions in Scotland are divided between large regional councils and smaller district councils. A similar system obtains in England and Wales. So a teacher working for a region can be the elected planning committee chairman of a district—and a district council truck driver can be leader of the region. Not so in the islands, where the council does everything but run the fire brigade and the police, these being controlled by joint boards.

In Shetland this ban removes over 2,000 full-time and part-time council employees from public life—including all of the teaching profession which, traditionally, has supplied some outspoken and generally radical-minded councillors throughout Scotland. Because the council is so involved in the economy, thousands more are effectively prevented from standing for election because they or their employer are subsidised by or do business with the council. There would be frequent and unacceptable conflicts of interest. Others are deterred because employers cannot or will not allow them enough time off work to attend the almost daily meetings of the council and its committees—all of which are held during working hours.

As a result, the pool of potential councillors is reduced to a puddle. Many seats are uncontested, the most common excuse of those refusing nomination being that they cannot spare the time to do the job. Almost two-thirds of the members returned for the 1986–1990 council were over retirement age, a figure wildly out of proportion to the number of pensioners in Shetland. Not surprisingly, there tended to be a conformity of interests and attitudes among these members which could make council meetings embarrassingly bland, ill-informed, confused, disorganised and meandering. The standard of chairmanship was sometimes pitifully low. Fortunately for public order, few citizens ever attended the meetings. If they had, there might have been rioting in the streets.

The temptation for a local journalist is to print verbatim reports of the proceedings of such meetings, to show the public how bad things really are. Sometimes, mischievously, we have succumbed to this temptation but supplies of newsprint are limited and subconsciously, perhaps, we reason that the public's image of the council is already bad enough. So we tend to summarise their ramblings into something like connected, logical order and translate their garbled utterances into something like plain, grammatical syntax. I often wonder if we are really serving the interests of democracy by tidying up reality in this way. But when you have known some of the members personally for so many years and have to live with them in the same community, it is a hard business to expose them to the full glare of public scrutiny. It may have to come, but we always live in hope that the next election will throw up a slightly better lot.

Some of the most original councillors, however, such as the venerable pensioner, Dr Mortimer Manson, who canvassed his Lerwick Clickimin constituency on foot in his first election when well into his seventies, have made useful contributions. As an octogenarian and determined pedestrian, he made an ideal chairman of the leisure and recreation committee, being free from all charges of vested interest in sports centres and other facilities for young folk.

Those councillors who are not retired must do the work in their spare time but, because of the council's extraordinarily wide range of powers and responsibilities, it has long since become a full-time job. So, powerful and knowledgeable officials tend to dominate and manipulate decision-making, which may or may not be a good thing, depending on your view of the elected members.

Because of the inadequacies of the councillors, the struggle with the oil industry has often been unequal—as evidenced by the curiously low profile of the public's representatives on the Sullom Voe Association, and by the fact that a four year legal action in pursuit of land rents of £100 million a year for the public's property at Sullom Voe ended in November 1988 with a deal which gave the council only a tenth of that sum. Needless to say, it was hailed as a "splendid" deal by the council. So splendid was it, indeed, that they refused to admit reporters to the signing ceremony although, quite bizarrely, they did let in a press photographer (the reason was that it was not a finished deal they were signing but a deal which still had some work to be done on it). The posed picture did not reveal details of the documents which a reporter's eye might have spotted. The full text was not revealed for another 11 months and, when it did come out, it showed that the councillors had had to give up some cherished financial yearnings.

Not all the problems of local democracy in Shetland are the council's fault. Despite the special powers of the ZCC Act, Shetland's status as a most-purpose local authority, and the oil income which makes possible the council's intervention in the local economy, councillors are still subject to central government restrictions on how they spend their own money, as are all other councils in Britain. Monetarist theories about controlling the money supply may have bitten the dust in Whitehall, but they are still applied in a fairly crude fashion to local government, even when spending oil funds which have nothing to do with the national exchequer. Until recently the Government even dictated how much could be spent under each of the council's budget headings. The total council budget, not just the portion of it which comes from central government taxation, is still subject to "spending consents", controlling the council's own money as well as what the Government gives it.

So there is still something rotten in the state of Shetland, despite the undoubted successes and benefits achieved by a local government with petro-pounds galore to spend. The solution, according to several outspoken public figures and the 15 percent of the local electorate who support them at elections, is even more self-government for Shetland. This is the policy of the Shetland Movement party, which has half a dozen councillors (and some supporters) in the council chamber. Others, myself included, think it would be a good idea to make the existing system of limited home rule work a bit better before we go any further. And we suspect that it may be no bad thing for central government to intervene from time to time to curb the wilder follies of a local government which controls everything from public lavatories and leisure centres to salmon farm licensing and the port of Sullom Voe. The trouble is that the present Government, which in Scotland is a junta repeatedly rejected by the electorate, cannot be relied upon to act wisely and in the public interest.

It is unlikely that the Shetland Movement and its demands for "more autonomy", falling short of home rule, would ever have come about without the oil money. Before oil, Shetland councillors were more often seen as fawning supplicants at the Government soup kitchen—bawling the odds, certainly; prophesying wholesale depopulation if the building of such and such a pier or road were not subsidised; and even being extremely rude about the persons and moral fibre of Scottish government ministers; but basically they were on their knees. The Government had the cash. Shetland didn't.

In contrast, the 1980s have brought us the spectacle of a local government whose self-assurance is much admired during its regular visits to Edinburgh and beyond to demand central government funds for its projects. At one point, hilarious in retrospect, the Shetland Movement even drew up a "budget" for a wholly independent Shetland. It was hastily withdrawn when a brief re-examination of its arithmetic showed that Shetland was heavily dependent on subsidies from the south. We were still mendicants, but we now had some means of our own. All this was heady stuff for our councillors, some of whom began to adopt the veneer of *gravitas* which, they imagined, befitted their new role as local statesmen treating with a powerful neighbour, the British state. Several bought new suits and neckties.

The late, and much lamented, Alexander Irvine Tulloch, in particular, acted the statesman for all he was worth, employing his naturally patrician airs to great effect when he was leader of the council. He often brought home the loot from Edinburgh for a favourite pier, school or road improvement. "A.I.", as he was universally known, often left Government ministers and civil servants in a state of awed astonishment, tinged with amused respect. As one of them once confided to a journalist friend of mine: "Mr Tulloch is the only council leader I know who flies in a chartered aircraft to London, drives in state to Dover House [the London base of the Scottish Office] and grants an audience to the Secretary of State for Scotland". He did, too. And he could also turn a frosty smile on those oilmen who displeased him—without, alas, persuading *them* to part with much of the loot.

The late Alexander Tulloch ("A.I.") in determined mood before a big meeting with the oil industry.
Photo: J. Wills Archive (photographer unknown)

Impatience with central government bureaucracy had as much to do with the origins of the Shetland Movement in the late 1970s as did the resurgent spirit of self-confidence in the Shetland middle class who contributed most of the movement's founding fathers—few mothers had anything to do with it. There was also a certain amount of envy involved, as people in Shetland looked to their neighbours in Faroe, to the Channel Islands, the Isle of Man and even the Aland Islands in the Baltic—and made unfavourable comparisons with the state of affairs at home.

Government restrictions on council spending were much resented; and there was a nagging fear, not entirely groundless, that the Government was planning to swindle Shetland out of its oil money—that is what happened, in effect, when for eight years of the 1980s, about £10 million a year in Government grants were withheld from the Shetland council because of the extra "resources" it enjoyed because of oil. Indignation was also rife, and justified, when Government-inspired financial concessions to business, on local property taxes, were extended to Shetland, severely damaging the council's tax take from Sullom Voe. The laws had been designed by well-intentioned Parliamentarians to stimulate industrial development in areas of high unemployment—hardly appropriate in Shetland with its low unemployment and overheated local economy.

British planning laws and building regulations also seemed inappropriate to Shetland's extreme climate—the large size of windows demanded by law was a frequent example quoted, although it is noticeable that the Norwegian-style luxury chalets, now in vogue with the Shetland Movement's supporters among the micro-bourgeoisie, tend to have gigantic triple-glazed windows. However, there were certainly some extremely silly regulations which contributed to the inappropriate high-density layout of some of Shetland's oil-age council housing schemes. Their structural and social problems have cost the council millions.

Another gripe was that the British Government had persistently refused to subsidise the Aberdeen-Lerwick ferry route as if it were a trunk road—giving rise to a lengthy period of moaning about the "road-equivalent tariff"—which ended only when the Treasury did agree to a fairly generous annual subsidy to P&O Ferries, the main shipping line on the route. On a smaller scale, Shetland Movement spokesmen (I cannot recall a woman making a public statement on their behalf) highlighted petty government restrictions which meant that although parts of Lerwick had a road safety problem, the council was not allowed to install "speed bumps" or roundabouts—that peculiarly British road layout which probably causes more accidents than anything else.

British laws on nature conservation were sometimes bitterly resented by crofters who were determined to shoot rare birds such as the great skua, popularly believed to attack their lambs. There was resentment of southern-born, university-educated scientists coming to Shetland to tell the locals how to look after "their" wildlife—something that the locals thought, often with little justification, that they could manage very well themselves, thank you. The resentment spilled over into diatribes against the harmless folk, mostly incomers, who made up the membership of the Shetland Bird Club and supported the Royal Society for the Protection of Birds, an outspoken defender of the Shetland environment—long before such sentiments became fashionable.

The founders of the Shetland Movement, men like John Graham, Jim Tait and John Goodlad, had impeccable liberal credentials and, if they ever harboured any xenophobic thoughts, kept them to themselves. But the movement's support came in part from less refined and thoughtful members of society who were attracted by simplistic ideas about Shetland for the Shetlanders and imagined that self-government meant telling soothmoothers to mind their own business and/or go to hell—the rednecks, in other words.

Despite the efforts of the movement's leaders to stress that their policies were designed for the benefit of everyone who lived in Shetland, irrespective of what their birth certificates said, the xenophobia of many of their supporters was palpable, unpleasant and undoubtedly led to ill-feeling between incomers and locals, leaving half-breeds like me and several thousand others not knowing where we stood at all.

In the second half of the 1980s it was the environmental issue which caused the greatest furore in Shetland, as in many other places. Hardy crofting sons of toil on the soil objected to soothmoothers declaring their land as nature reserves and sites of special scientific interest, and forbidding them to plough up rough grazing or to drain bogs—homes for rare birds (rare in British terms) such as the whimbrel and the red-necked phalarope. This was particularly pronounced in the island of Fetlar where some fairly violent language was used until Dr Mike Richardson, the pleasant, commonsensical scientist in charge of the Nature Conservancy Council in Shetland, eventually persuaded crofters that they could cash in on their rare rough grazing and birdy bogs by entering "management agreements" in return for what amounted to pensions from the Government. This was to compensate them for not taking part in the council's multi-million pound agricultural development plan, which history may regard as the greatest environmental disaster in Shetland since the introduction of the all-devouring sheep. The conservation deals in Fetlar and elsewhere cost a lot of taxpayers' money but were a sensible solution—the Shetland equivalent of being paid not to grow alfalfa.

Another cause of resentment of Big Government was the deepening crisis of a fishing industry whose technology had outgrown the natural resources upon which the boats, and hundreds of Shetland jobs, depended. Before Britain joined the European Economic Community in 1973 there were national fishing limits—boats over a certain size could not fish within three miles of the shore and most foreigners were kept outside twelve miles, except by agreement. But within ten years of entering Europe, Shetland's rich fishing grounds were wide open. Boats from all over Europe could legally fish up to the beaches. An essential refuge for the young of the commercial white fish stocks was destroyed, with entirely predictable results. Unenforceable catch quotas replaced the old geographical limits and were found wanting as a method of controlling over-fishing.

By 1986 European and British fishery regulations seemed hopelessly irrelevant to the needs of the Shetland fleet, which caught and still catches only about 10 percent of the fish taken in Shetland waters. Demands for local control of how much fish should be taken, with which types of gear, and when and where the catch could be landed formed the central plank of the Shetland Movement's programme.

The movement was started by a group of middle class intellectuals in Lerwick—mostly teachers, factory owners and managers. Many of them were renegades from the Shetland Labour Party, although a few tried and failed to hold dual membership for a while. They were the sort of people who, anywhere else in Britain, would have left Labour to join the Social Democratic Party in 1979—and it is no coincidence that no SDP branch was ever established in Shetland (the Liberals were too strong, for another thing). They began as a discussion group where public-spirited individuals, dissatisfied with what the established political parties were offering Shetland, set up a popular forum to debate solutions to what they perceived as a crisis. I must admit to having sympathised with them—until their wilder *folies de grandeur* became too apparent.

They moved on to become an organised group, not yet contesting elections but questioning party candidates on their policies for local self-government and the issues which I mentioned above. Not surprisingly, this led them to endorse the Scottish National Party candidate in the 1983 election, who came third with about 15 percent of the vote in the Orkney and Shetland constituency (Orkney had its own Movement, a minuscule affair composed largely of embarrassing eccentrics and political flat-earthers).

By the time of the 1986 local government elections, the movement were putting up candidates of their own—and managed to get eight elected, compared with the dozen or so in the 1982–86 council who had professed support for the movement's aims, while maintaining their independent status. That marked the peak of the party's fortunes: their councillors failed to act as a co-ordinated group on the council; senior movement figures such as Councillor John Graham became associated with the disappointing deal with the oilmen over the Sullom Voe rent; and public interest waned. The movement's first venture into British parliamentary politics was in June 1987. John Goodlad, the secretary of the Shetland Fishermen's Association, stood as the joint Orkney and Shetland Movement candidate. It was a disappointment. Despite the Scottish Nationalists standing down to give him a clear run, he ended up third, with a slightly lower share of the poll than the nationalists had managed in 1983. By 1987 the local electorate had seen a large slice of home rule in action, and it was not ready for any more just yet.

Even so, the movement could claim to have put the issue of self-government on the agenda, something that would have been unthinkable back in 1972. Its agitation, at the time of the stillborn attempt to give Scotland a parliamentary assembly of its own in 1979, had led to a Government-appointed inquiry, the Montgomery Commission, which looked at options for the future government of the Northern and Western Isles of Scotland. The commission came up with suggestions for tinkering with the existing system of local government, most of which were ignored by the increasingly centrist Government of Margaret Thatcher. And it entirely rejected the arguments of the more ardent Movementeers for something like Faroese status—home rule in everything but foreign affairs and defence with a guaranteed subsidy from central funds—Danish funds in the Faroese case. No-one took to the streets of Lerwick to demonstrate against this rebuff, although the indignation in the letters columns of the local press went on for years afterwards.

The Shetland Movement's problem was that it had talked too much. Most people were doing very nicely, thank you, and had become bored stiff with the endless monologues about the need for constitutional change. The movement still has no charismatic public speaker or pamphleteer, its utterances being characterised by complex and tortuous language of Dickensian pomposity—Mr Bumble would have loved them.

It was not just the lack of style and communications flair—many of its criticisms of the inappropriateness of central Government's policies could apply equally to anywhere in Britain—particularly when Mrs Thatcher began her determined assault on the National Health Service, public education, the Welfare State and the public ownership of gas, water, electricity and the major transport utilities. On these issues, the great political arguments of the 1980s in Britain, including Shetland, the movement was rather quiet.

A further problem locally was that many of the movement's policy pronouncements were so bland that almost anyone could agree with them. For example, no Labour, Liberal, Green or Nationalist voter in Shetland would disagree with the proposition that the conservation of fish stocks should be based on local control—although some, looking at the overfishing in Faroe where just such a system exists, might have doubts about the wisdom of leaving it all to the fishermen; nor would they disagree that Shetland's oil income should be left alone by the Government; or that public transport in remote communities should be subsidised. And a key movement policy—for full-time, paid councillors, had been Labour and Nationalist policy for many years.

Here we are back at the root of Shetland's chronic political crisis—the shortage of candidates for council seats. We have seen how thousands of islanders are effectively prevented from standing for election, although I do not suggest that more than a small fraction of them would wish to, even if there were no obstacles. The system militates against all council employees, wage-earners, home-makers and small business owners. If councillors could be paid, say, the average industrial wage, plus necessary travel and office expenses, there would immediately be more competition at election time. Being full-time, they could do much of the supervisory and planning work at present left to an army of officials, assuming as I do that making membership of the council a paid job would attract a higher calibre of candidate. If there were also a grant to tide a councillor over for a few months, if he or she lost their seat and could not immediately return to other gainful employment, even more people would be tempted to stand at elections. But no-one is going to hazard their career, their business or the welfare of their family for the public good, if being elected means being on the breadline. Also, existing attendance allowances, laid down by national laws, are not generous. Even with the number of meetings the Shetland council holds, the allowances do not make up a living wage.

No-one any longer suggests that British MPs should be gifted amateurs with private incomes. They are paid to work full-time—although many Conservative members do not, and hold down lucrative positions in business and commerce at the same time. A council committee chairperson in Shetland has far more power over public money, and makes decisions which affect more voters, than the average back-bench MP. Some would say that such committee leaders have more power than a junior Government minister. What our councillors do matters, immediately. What MPs do is of little consequence, mostly, under our system of elected dictatorship where Prime Ministers have untramelled powers, far in excess of those exercised by Presidents of the United States. But we still expect our councillors to muddle along somehow, as more or less gifted amateurs, even when they are dealing with global organisations like the oil industry. The marvel is that so much has been achieved in Shetland under such a system.

Home rule is a nice idea, but it is no use unless it is accompanied by a large dose of open democracy. Home rule by an unaccountable, secretive cabal is not progress. We are often told that small is beautiful. My bitter experience is that small can be pretty ugly.

The government of Britain is a notoriously secretive and class-ridden business. Draconian laws on trade unions and new restrictions on access to official information have been introduced since 1979 by a Government which has selected once-powerful minorities, be they dockers or doctors, miners or maths teachers, and systematically crushed their political clout (and often civil rights). Visitors from North America are usually astonished at the dictatorial tone and methods of our Government in this self-styled "Mother of Parliaments". And they cannot understand the obsessive secrecy which surrounds the work of most government departments.

They were also surprised, to say the least, when they arrived in Shetland and found that councillors, including three Labour ones, whose party favours a Freedom of Information Act, had resorted to the courts to suppress reports in the local press about secret negotiations between the council and the oil industry. That is what happened in August 1988. In that month I had great difficulty in persuading a visiting US TV producer that this was actually happening. His incredulous film crew spent much of the evening discussing it over the dinner table.

The issue arose when a contact of mine inadvertently allowed me to have sight of certain documents, referring to a secret briefing between a small group of senior councillors and their Edinburgh lawyers in May 1988. Details were subsequently confirmed by other sources. The subject of the meeting was how they could bring pressure to bear on the oil companies to settle out of court on that ticklish business of the Sullom Voe rent. The councillors, it appeared, had been led to believe that their ultimate sanction was to withdraw BP's permission to occupy the site. But their new QC (Queen's Counsel), appointed after his predecessor had been made a judge, cannily pointed out the unpalatable truth: they could not use this sanction because they had already agreed, back in November 1978, to let BP and partners occupy the council's land at Sullom

Voe for as long as it might take to reach a rent deal. There was no time limit. There should have been. They were stuck in a bind of the council's own making, and 12 of the 25 serving members had been councillors when this foolish agreement was made. All very embarrassing—they would have to take what BP offered, or go on through the courts for years, perhaps until 1994, spending more millions to have it all sorted out.

The oil industry, having even more expensive lawyers at its disposal, had noticed this 10 years earlier and had frequently referred to it when briefing journalists about the progress of the interminable rent negotiations. "Just a piece of paper" was what they called the agreement to agree upon a rent.

The Shetland Times published the revelation that the council had been told this, and also forecast fairly accurately what the rent would be when the deal was eventually struck, five months later. A secret council meeting ensued at which, our informants quickly told us, they had decided to take *The Shetland Times* to court. There was just time to lodge our defences in the Court of Session at Edinburgh before the writ arrived. An interim interdict was granted by the puzzled judge and there we were—forbidden to repeat what we had already printed; and threatened with having to reveal our sources, return any copies of the documents in our possession and being required to give an undertaking not to reveal any more of the council's confidences.

This was, as anyone with the slightest acquaintance of the law of Scotland will know, preposterous. There may be laws requiring two parties to a piece of confidential information not to divulge it to anyone else but, once it has been divulged, there is not and cannot be any duty of confidence upon a third party such as *The Shetland Times*. The less foolish councillors knew this and one of them, a Labour member with a law degree, admitted as much when I challenged him in the street: "We knew it would shut you up for up to two years, and by then it will all be settled", he told me, referring to the inordinate length of time such actions can take in Scotland's congested legal system.

It did not entirely shut us up, of course; it boosted sales, as extra readers queued to buy extra copies of our editions with blank spaces in them. And, by a careful use of cartoons and disguised "fairy story" pieces in our diary column, we were able to tell the public some of what the councillors had wanted to suppress.

One of the things they wanted to suppress was that the council leader had expressed his concern at the loss of "credibility" which would follow the revelation that their only sanction against BP was worthless; and, of course, their wish to preserve their former QC, now on the bench, from the embarrassment of the revelation that his successor disagreed with his opinions. They also wanted to avoid publication of all sorts of lawyers' and chartered surveyors' estimates of the likely outcome of the case—which were written in such convoluted language that no-one, certainly not me, could ever have understood them.

It was a foolish miscalculation by the council. Even more foolish was the several thousand pounds they then spent on a "mole-hunter"—a retired policeman who was brought up to Shetland to take statements and interview the suspects—quizzing them on whether they had ever lunched or partied with me and asking what they were up to at two in the morning on a certain day in July 1988. If you had dreamt it all up for a Christmas pantomime, people would have criticised the script as being too fanciful.

It kept the paper "censored at the instigation of Shetland Islands Council", as our front page banner proclaimed, for seven months, until the mounting legal expenses (which eventually came to some £2,600 on our side) and public disapproval brought the councillors to their senses and they dropped the case. Had it come to court, I would have faced a heavy fine for refusing to reveal my sources (plural). The idea seemed to be that newspapers should be penalised by the law for doing their job, but councillors and officials who broke the law by accepting BP trips ("reward other than their proper remuneration", as the Local Government (Scotland) Act 1973 puts it so elegantly) were to be let off with a light tap on the wrist.

In retrospect, the Court of Session interdict was an example of over-reaction by bigwigs who genuinely believed that they had the interests of Shetland at heart—and believed that the duty of the local press was to act in concert with them against the perceived "enemy within", as one of our councillor censors had once described BP. They pretended that BP did not know of the fatal flaw in their argument and also that publication of it would affect the inevitable outcome of the rent negotiations.

The duty of the press, of course, was and is to stay clear of such political entanglements and to tell its readers what it knows, leaving the interpretation of the facts to its opinion columns where, incidentally, we had consistently supported the council's demand for a fair rent from BP. One of the funniest twists to the story was the allegation in *The New Shetlander* magazine, co-edited by one of the councillors, that (for some unexplained reason) the local newspaper and its then editor had suddenly become traitors in the pay of BP. I did not bother to sue. More amusement was to be had when a councillor became outraged that we did not print his letter implying that I had stolen the famous documents. I do not print defamation, even of myself, although suing myself for being party to a libel might have had its entertaining moments, even if *veritas* could not have been a defence in this case.

Yes, it was all rather ridiculous, but there was also that important legal principle on the duty of confidentiality. Ours was a test case, my paper's proprietor and I were surprised to learn. The only parallel case that our QC could find (when he had stopped laughing at the absurdity of the council's pretensions) involved the British Government's attempts to stop *The Scotsman* newspaper reprinting extracts from the memoirs of one Anthony Cavendish. He was a long-retired, minor espionage agent who wished to clear his former boss, the late Maurice Oldfield, of what he regarded as a "traitor" smear. *The Scotsman* had a little more money than *The Shetland Times* and fought it all the way to the House of Lords. Their Lordships were of the same opinion as our QC, and Mr Cavendish's sad memoir was duly published, some months after our own little local difficulty.

The incident revived my interest in the long and so far unsuccessful campaign for a Freedom of Information Act in Britain, modelled on that in the United States. And it taught me that, if it ever comes, it will need to cover local government as well, or small may become even uglier.

Despite their occasional eccentricities, Shetland's councillors are by no means the most secretive local authority in Britain—if only because in a place this small it is difficult to keep anything secret for long. There are many other councils, notably those dominated for many years by a single political party, where concealment, backroom fixing and stealthy lobbying are the normal way of doing business. At least there are no party caucuses in Shetland. And most councils in Scotland make copious use of the two Acts of Parliament which allow them to restrict information—the Local Government (Scotland) Act 1973 and the Local Government (Access to Information) Act 1985. The latter was hailed by Mrs Thatcher, who put it through Parliament, as a milestone in making local government more open and accountable. It has not worked out like that in Shetland, nor in the only other Scottish councils with similarly wide powers—the Western Isles and Orkney.

These laws are complicated, but basically they say that a council cannot make public anything it has been told in confidence by the Government—that can apply even to the advice of consultants on an agricultural development program. More understandably, it must not discuss personnel matters, such as staff interviews, appointments and dismissals, in public, nor publish background papers for such items; it must not reveal details of unfortunates who apply for social grants or who have their domestic affairs dealt with by the council's social work department, for example.

Much of this is sensible and necessary confidentiality, as long as public business is being conducted by competent professionals, scrutinised by our elected representatives. It is no business of the public's to know which disabled pensioner is having a grant for a handicapped person's car or improvements to his or her kitchen. We do not need to know or publish the names of children being taken into care. The social background of students applying for college grants is properly private. Nor should we be interested in the references provided by candidates for teaching jobs or refuse-truck drivers.

All this is commonsense, if the personal lives of innocent private citizens are to be protected from gossip and sensational publicity. In a place the size of Shetland, quite enough information is already in the public domain about people's private lives.

Shetland Islands Council, and its counterpart in Orkney, has gone further. A whole lot further. Under the law, there are many categories of information described as "exempt". Councillors are empowered to decide for themselves if some information under these categories should be kept secret. Very often, they use the "exempt" clauses of the Acts to discuss in secret serious foul-ups committed by their staff or by themselves. Parliament (and even Mrs Thatcher) did not intend the law to be used to protect the incompetent.

In a recent example, a major miscalculation of the number of students likely to apply for college grants, and the consequent shortage of funds in the education budget, which left 70 students stranded with no money, was discussed in secret session on the pretext that individual cases were being considered. In fact, the meeting did not discuss individual cases; it discussed policy and budget matters which should have been in public. Of course, reporters did not want the details of the financial and family circumstances of the individual applicants—but these could only be discussed after the debate about the defective budget had been completed. You cannot discuss how to spend a budget until you have decided what it is to be. Eventually, after a public outcry, another session was held, this time in public, where a report by a finance official sought to apportion blame but concluded that it was all the fault of the system, rather than any individual's failings. Cash was transferred from another budget, and most of those students who were eligible got their money, although by that time some had given up the college places for which they had qualified.

It is not just when something goes wrong that the doors are closed: we are not allowed to know the details of tenders by private companies for public business, on the grounds that this might expose confidential company finances and lead to a cartel of bidders, to the detriment of the public. This is nonsense—in any small town all the bidders know each other anyway, and there are informal cartels all over the place. Exposure of tenders would tend to break up cartels, increase competition and give the public better value for money. If companies do not like publication of their contract tenders, they can always look for private work where confidentiality is guaranteed. But what is done with the public's money ought to be public.

Very large subsidies from council funds are handed out in secret. Some, such as loans of up to £400,000 for fishing boats are belatedly published, in very small print, in the council's annual reports and accounts, which have been as much as four years late. Some recipients, laughably in a community of this size, are referred to only by their initials. But most cases are kept secret, unless the recipient requests or agrees to publicity. Factories, airlines, bus operators, fishing boats, fish farms, farms and crofts, boarding houses, hotels, shops, wholesalers and businesses of all descriptions receive grants and loans worth in total several million pounds a year—mostly in secret. In the United States, in Scandinavia and in many other western democracies this would not be tolerated. In Shetland it is normal.

Granted, all this is small beer, parish pump stuff. And every small town journalist in the world has a bind about the local worthies who see their job as concealing what he would like to be revealing. And every reporter worth his salt has his moles, who send him plain brown envelopes containing secret documents, and tip him off with anonymous phone calls. Yes, so do I, but getting information out the back door has its dangers—not just the danger of court action. Moles can sometimes be people with axes to grind; their information may not always be accurate and is usually incomplete. There will always be moles but we should not have to rely on them to this extent. Most of what they send my way should be coming out of the front door of the Town Hall and the Sullom Voe terminal.

What was certainly not parish pump stuff was the secrecy about the council's deals with the oil industry. These involved hundreds of millions of pounds. There is no doubt that most Shetland councillors grew as weary of this issue as I was. I grew so weary of it by August 1989, indeed, that I referred the whole sorry business to the local government Ombudsman in Scotland, to see if he could sort it out—leaving me free to get on with the infinitely happier task of editing sports reports and writing book reviews. His reply was an education: he advised me that I should take them to court if I thought they had broken the rules; that no council was under any legal obligation to give details of why it went into secret session; and that whether or not minutes of secret meetings were sufficiently informative was a matter of "subjective opinion". This depressing admission of bureaucratic impotence, from the man who is supposed to be a watchdog on local authorities, will be useful ammunition in the campaign for a local Freedom of Information Act.

Public pressure, not official watchdogs, has now forced the Shetland council to publish the text of their deals with BP and friends. The arguments about "commercial confidentiality" were abandoned in the end, but, for fifteen years, the excuse that the oilmen would not like all this to be public ensured that ordinary citizens were denied access to details of what was being done in their name. And we still do not know what has really gone on in the Sullom Voe Association. Its minutes, if they are ever published, are hardly likely to be informative, to judge by the standard of record-keeping (and of English syntax) in other committees connected with the council.

Even in Britain, national governments eventually have to publish their records. There is a thirty-year rule, which keeps records of the British cabinet secret until the fuss has died down; and a one hundred-year rule, which keeps the lid on the secret service stuff and politically embarrassing material. Many strange things, such as the Scottish police files on informers during the General Strike of 1926, are still secret but in the end the Government will have to publish even those.

Local government is different. There is no thirty-year rule and not even a hundred-year rule. The inside story of the council's negotiations with BP can be kept secret until Doomsday and beyond. Worse still, quite minor officials have the power to "weed" council records and to shred or burn material which they judge to be no longer of interest or value. The question is—of interest or value to whom?

Well, to the writers of history books, if to no-one else. If future historians ever do see the "record copy" of the council's papers in the SIC Department of Administration and Legal Affairs, they may well find that the history of Shetland's oil era has been edited. We shall perhaps know more about what went on in that court at Scalloway Castle, over 370 years ago, lit by the flame of a seal-oil lamp.

Unchanged by it all: the peaceful harbour of Burravoe, Yell.
Photo: Jonathan Wills

Index

ISER BOOKS

Studies

41 **A Place in the Sun: Shetland and Oil–Myths and Realities** —Jonathan Wills
40 **The Native Game: Settler Perceptions of Indian/Settler Relations in Central Labrador**—Evelyn Plaice
39 **The Northern Route: An Ethnography of Refugee Experiences** —Lisa Gilad
38 **Hostage to Fortune: Bantry Bay and the Encounter with Gulf Oil**—Chris Eipper
37 **Language and Poverty: The Persistence of Scottish Gaelic in Eastern Canada**—Gilbert Foster
36 **A Public Nuisance: A History of the Mummers Troupe**—Chris Brookes
35 **Listen While I Tell You: A Story of the Jews of St. John's, Newfoundland**—Alison Kahn
34 **Talking Violence: An Anthropological Interpretation of Conversation in the City**—Nigel Rapport
33 **"To Each His Own": William Coaker and the Fishermen's Protective Union in Newfoundland Politics, 1908–1925**—Ian D.H. McDonald, edited by J.K Hiller
32 **Sea Change: A Shetland Society, 1970–79**—Reginald Byron
31 **From Traps to Draggers: Domestic Commodity Production in Northwest Newfoundland, 1850–1982**—Peter Sinclair
30 **The Challenge of Oil: Newfoundland's Quest for Controlled Development**—J.D. House
29 **Sons and Seals: A Voyage to the Ice**—Guy Wright
28 **Blood and Nerves: An Ethnographic Focus on Menopause**—Dona Lee Davis
27 **Holding the Line: Ethnic Boundaries in a Northern Labrador Community**—John Kennedy
26 **'Power Begins at the Cod End': The Newfoundland Trawlermen's Strike, 1974–75**—David Macdonald
25 **Terranova: The Ethos and Luck of Deep-Sea Fishermen**—Joseba Zulaika (in Canada only)
24 **"Bloody Decks and a Bumper Crop": The Rhetoric of Sealing Counter-Protest**—Cynthia Lamson
23 **Bringing Home Animals: Religious Ideology and Mode of Production of the Mistassini Cree Hunters**—Adrian Tanner (in Canada only)
22 **Bureaucracy and World View: Studies in the Logic of Official Interpretation**—Don Handelman and Elliott Leyton
21 **If You Don't Be Good: Verbal Social Control in Newfoundland** —John Widdowson
20 **You Never Know What They Might Do: Mental Illness in Outport Newfoundland**—Paul S. Dinham

Papers

14 **Indigenous Peoples and the Nation-State: Fourth World Politics in Canada, Australia and Norway**—Noel Dyck (ed.)
13 **Minorities and Mother Country Imagery**—Gerald Gold (ed.)
12 **The Politics of Indianness: Case Studies of Native Ethnopolitics in Canada**—Adrian Tanner (ed.)
11 **Belonging: Identity and Social Organisation in British Rural Cultures**—Anthony P. Cohen (ed.) (in North America only)
10 **Politically Speaking: Cross-Cultural Studies of Rhetoric**—Robert Paine (ed.)
9 **A House Divided? Anthropological Studies of Factionalism**—M. Silverman and R.F. Salisbury (eds.)
8 **The Peopling of Newfoundland: Essays in Historical Geography**—John J. Mannion (ed.)
7 **The White Arctic: Anthropological Essays on Tutelage and Ethnicity**—Robert Paine (ed.)
6 **Consequences of Offshore Oil and Gas—Norway, Scotland and Newfoundland**—M.J. Scarlett (ed.)
5 **North Atlantic Fishermen: Anthropological Essays on Modern Fishing**—Raoul Andersen and Cato Wadel (eds.)
4 **Intermediate Adaptation in Newfoundland and the Arctic: A Strategy of Social and Economic Development**—Milton M.R. Freeman (ed.)
3 **The Compact: Selected Dimensions of Friendship**—Elliott Leyton (ed.)
2 **Patrons and Brokers in the East Arctic**—Robert Paine (ed.)
1 **Viewpoints on Communities in Crisis**—Michael L. Skolnik (ed.)

Mailing Address:
ISER Books (Institute of Social and Economic Research)
Memorial University of Newfoundland
St. John's, Newfoundland, Canada, A1C 5S7